T0183012

CAMBRIDGE LIBRARY COLLECTION

Books of enduring scholarly value

Polar Exploration

This series includes accounts, by eye-witnesses and contemporaries, of early expeditions to the Arctic and the Antarctic. Huge resources were invested in such endeavours, particularly the search for the North-West Passage, which, if successful, promised enormous strategic and commercial rewards. Cartographers and scientists travelled with many of the expeditions, and their work made important contributions to earth sciences, climatology, botany and zoology. They also brought back anthropological information about the indigenous peoples of the Arctic region and the southern fringes of the American continent. The series further includes dramatic and poignant accounts of the harsh realities of working in extreme conditions and utter isolation in bygone centuries.

An Account of a Geographical and Astronomical Expedition to the Northern Parts of Russia

Impressed by the discoveries of Captain Cook, and conscious that Russia was lagging behind other countries in terms of navigation and exploration, Catherine the Great commissioned an expedition in 1785 to chart the coastline in the far north-east of her empire. Born in Middlesex, Joseph Billings (1758–1806) had sailed under Cook but entered Russian service in 1783. He was chosen to lead the expedition, which would last for nine years. Written up by Martin Sauer, secretary and translator to the expedition, this illustrated account was first published in English in 1802, documenting the sheer scale of the task and the range of scientific activities carried out. Notable for producing the first accurate maps of the shoreline and islands of east Siberia, the expedition also contributed to the ethnographic and zoological knowledge of this most inhospitable of environments.

An Account of a Geographical and Astronomical Expedition to the Northern Parts of Russia

MARTIN SAUER

CAMBRIDGE
UNIVERSITY PRESS

CAMBRIDGE
UNIVERSITY PRESS

University Printing House, Cambridge, CB2 8BS, United Kingdom

Published in the United States of America by Cambridge University Press, New York

Cambridge University Press is part of the University of Cambridge.
It furthers the University's mission by disseminating knowledge in the pursuit of
education, learning and research at the highest international levels of excellence.

www.cambridge.org
Information on this title: www.cambridge.org/9781108066655

© in this compilation Cambridge University Press 2013

This edition first published 1802
This digitally printed version 2013

ISBN 978-1-108-06665-5 Paperback

AN

ACCOUNT

OF A

GEOGRAPHICAL AND ASTRONOMICAL

EXPEDITION

TO THE

NORTHERN PARTS OF RUSSIA,

FOR ASCERTAINING THE DEGREES OF LATITUDE AND LONGITUDE OF
THE MOUTH OF THE RIVER KOVIMA;
OF THE WHOLE COAST OF THE TSHUTSKI, TO EAST CAPE;
AND OF THE ISLANDS IN THE EASTERN OCEAN, STRETCHING TO
THE AMERICAN COAST.

PERFORMED,

By Command of Her Imperial Majefty *CATHERINE THE SECOND,*

EMPRESS OF ALL THE RUSSIAS,

BY COMMODORE JOSEPH BILLINGS,

In the Years 1785, *&c. to* 1794.

THE WHOLE NARRATED FROM THE ORIGINAL PAPERS,

BY MARTIN SAUER,

SECRETARY TO THE EXPEDITION.

LONDON:

Printed by A. Strahan, Printers Street;
FOR T. CADELL, JUN. AND W. DAVIES, IN THE STRAND.
1802.

TO

THE RIGHT HONOURABLE

SIR JOSEPH BANKS, BARONET,

A KNIGHT COMPANION OF THE MOST HONOURABLE ORDER OF THE BATH;

A MEMBER OF THE KING'S MOST HONOURABLE PRIVY COUNCIL;

PRESIDENT OF THE ROYAL SOCIETY,

&c. &c. &c.

THIS WORK

IS,

WITH GREAT DEFERENCE AND RESPECT,

INSCRIBED,

BY

HIS MUCH OBLIGED,

MOST GRATEFUL,

AND OBEDIENT SERVANT,

MARTIN SAUER.

PREFACE.

———

THE extraordinary difcoveries of the ever-memorable circum-navigator Cook infpired all Europe with an enthufiaftic defire of being acquainted with the parts of the globe ftill remaining un-known.   Ruffia, though more interefted in thefe events than any other Power, being engaged in different purfuits, did not confider the diftant and barren regions belonging to her own Empire as of fufficient importance to juftify the expence and trouble of ex-ploring them ; until the genius of the country was completely rouzed by the animating intelligence communicated in the " Ac-count of the Ruffian Difcoveries between Afia and America, by the Reverend William Coxe," which the late Catherine the Second commanded to be tranflated for her own perufal, al-though the original Papers were in the Archives of the Admiralty at St. Peterfburg *.

———

* I am happy to find, that the author has collected very confiderable materials for an enlarged edition of this work ; which it is hoped he will not long withhold from the public, as the book is out of print, and cannot be procured.

The

The Court of Ruffia was aftonifhed at the difcoveries already made, by its own roving fubjects, of iflands, and of a continent, of which latter it had indeed an idea, but not the fmalleft notion of its extent or proximity to its own territories, and only fuppofed that it might be America. However, thefe voyagers did not afcertain the geographical fituations of places, nor explain the advantages that they offered to the country to which they belonged; nor, in fhort, any thing more than their mere exiftence.

The amazing extent of dominion acknowledging the fovereignty of Ruffia, independent of her late acquifitions by conqueft, became now the fafhionable topic of converfation at Court. Mr. Coxe, being at St. Peterfburg, took advantage of the favourable moment to fuggeft an Expedition, to complete the geographical knowledge of the moft diftant poffeffions of that Empire, and of fuch northern parts of the oppofite continent as Captain Cook could not poffibly afcertain. The learned Dr. Pallas, then in great favour, undertook to make the neceffary reprefentations to Her Imperial Majefty, who, well pleafed with the hint, immediately approved of the plan drawn out by thefe two Gentlemen; and Count Befborodko was, in confequence, ordered to prepare a Mandate for the Admiralty: this was in the autumn of the year 1784. Mr. Billings, who had juft received a Lieutenancy, faid, that he had been the

Aftro-

Aftronomer's Affiftant in Captain Cook's laft voyage; and he was therefore thought a proper perfon to conduct the enterprife.

Matters remained thus till the French Papers announced the departure of Count de la Peroufe, in July 1785, on a voyage of difcovery. Upon this, the undertaking was refumed with energy; and on the 8th Auguft following, an Ukaze, or Mandate, figned by the Emprefs, was fent to the Admiralty; on which were founded the INSTRUCTIONS to CAPTAIN BILLINGS, given in the APPENDIX to this Volume, No. V.

Every thing was procured that appeared likely to contribute to the fuccefs of the enterprife; every imaginable encouragement was awarded to all the officers and men; and orders were fent to the Governors, Commanders, &c. throughout the vaft extent of Siberia, to give all poffible affiftance.

Captain Billings had permiffion to felect his own officers, and to take fuch hands as he judged neceffary.

I was perfonally acquainted with Doctor Pallas and Mr. Billings, both of whom requefted that I would accompany the Expedition as Private Secretary and Tranflator; and, on receiving the

a   promife

promife of permiffion to publifh my remarks upon my return, I agreed.

On the 10th March 1794 I returned to St. Peterfburg, in a very critical ftate of health, which continued impaired during the whole of the following fummer, and induced Doctor Rogers (now in London), the Doctor (Merck) and Surgeon Major (Robeck) of our Expedition, to form an opinion, that the feverity of a Ruffian winter might prove of bad confequences; and they recommended my vifiting a milder climate for a fhort time. In confequence of this profeffional opinion, I petitioned Captain Billings, conformably to the eftablifhed rules of the fervice; requefting him to reprefent my fituation to the Admiralty, and to procure me leave of abfence for about four months. This was on the 2d of September 1794; and on the 5th of the fame month, TOWARD MIDNIGHT, *I received a very unexpected and unfavourable* ANSWER. It is not my intention, however, to enumerate hardfhips, or make a merit of fufferings; but to give the beft account I can of fuch occurrences as immediately concern the Expedition, and as appear to me moft likely to intereft my Readers.

In the mean time I embrace this opportunity of acknowledging my great obligations to the undermentioned Gentlemen, then
                                                              inhabi-

inhabitants of St. Peterſburg, for the particular marks of friendſhip which I received at their hands.

William Porter, Eſq.
Mr. William Jones.
Mr. Alexander Grant.
Mr. Laurence Brown.
Thomas Warre, Eſq.
William Wilſon, Eſq.
Alexander Shairp, Eſq.
John Booker, Eſq.
Doctor Simpſon.
Doctor Guthrie.
Mr. John Samuel Barnes.
— John Venning.
— William Glen Johnſton.
— John Glen Johnſton.
— Edward James Smith.

Upon my arrival in London, however, I experienced no leſs
generous treatment. M. Garthſhore, M.D. F.R.S. and A.S. has
my ſincere thanks for his protection; as alſo the Reverend
William Coxe, and the Reverend London King Pitt.

Thomas

Thomas Harvey, Efq. who particularly affifted me in Ruffia, has ftill heaped obligations upon me here; as have alfo Charles Grant, Efq. and Doctor Rogers.

My warmeft acknowledgments are likewife due to James Gibfon, Samuel Stratton, and John Rowlatt, Efqrs. for their friendfhip and recommendation.

The many kindneffes received from Mr. William Lotherington, and Mr. Edmund Rodd, my fellow-traveller from Ruffia, will remain indelible in my remembrance.

During my travels, I was frequently neceffitated to make notes on fmall pieces of paper; thofe I have faithfully tranfcribed; but in fome inftances I have been obliged to refer to memory; which circumftance, added to the obliterated ftate of feveral outlines traced with a black lead pencil, would have prevented my giving a chart of the two continents, had not Mr. Arrowfmith requefted to fee my remarks, which he compared with former difcoveries in thefe parts; and, obferving that the correfponding diftances (particularly Shalauroff's chart) agreed with Captain Billings's aftronomical obfervations in the Icy Sea, as did alfo the fketches of the natives, it plainly appeared to him, that he could venture to lay down the Shalatfkoi promontory, and the whole coaft between the eaftern promontory of Afia and the Kovima with

5   tolerable

tolerable exactnefs; which proves the general fault in the Ruffian charts, where the coaft is carried confiderably too far north. The fituation of the iflands between the two continents, as laid down in the chart, may be pronounced juft; but I feel myfelf infinitely obliged to Mr. Arrowfmith for the pains he has taken. I am equally fenfible of Mr. Alexander's merit in the judicious arrangement of the drawings and coftumes, which has enabled me to prefent the Engravings, exact in their refemblances, and exe-cuted in a manner highly pleafing to myfelf. While indulging my own fenfations in paying the tribute of refpect and gratitude to thofe who have befriended me, I ought not to overlook the kindnefs and liberality which I have experienced from my Pub-lifhers; but, as I am perfuaded that their behaviour to me is merely the ordinary courfe of their profeffional practice, I fhall reftrain my feelings, and avoid the rifk of offending them by being more particular.

Upon mature deliberation on the extent and tendency of this Work, I think it neceffary to call publicly on the Commander of the Expedition, and my brother officers, to correct any miftakes in my narrative *, or to elucidate fuch intricacies as may have

---

* My narrative of the voyages is taken from the journal written for Captain Billings, which I copied from the fhip's journal kept by the Mafter Batakoff and his mates. I am apprehenfive, that fome of the bearings are not perfectly correct; and I acknowledge that in many places I am not capable of faying whether the computed diftances are geographical or German miles; both meafures having been ufed by the original journalifts.

arifen

arifen from my want of knowledge in the different branches within the limits of their profeffional ftudies. My object has been to travel with my eyes open, and to relate what I have feen in the fimple language of truth.

Feb. 1802.

EXPLANA-

EXPLANATIONS

OF

Ruffian and other Foreign Words made ufe of in the following Work.

———————

BAIDAR; a term ufed at Ochotſk, Kamtſhatka, &c. for boats, whether large or fmall. They are pointed at both ends, and conſtructed as follows: A keel and three frames, the lower to form a flat-bottom, the fecond to fupport the thwarts, and the third to ferve for the gunnel; light knees and ground timbers are lafhed to the keel and the frames with whales' fins: The raw hides of fea-animals are drawn over, to ferve inftead of fheathing. They draw only a few inches water, carry a confiderable burthen, are excellent furf-boats, and very ufeful in coafting excurfions; as four men can carry one of them which admits of twelve rowers; at night they are turned keel upwards, and ferve inftead of tents. The fmaller are quite covered, leaving only one, two, or three openings for the rower.

BAZAR, or RENOK; a market for vegetables, hard and wooden ware, &c. Any perfon is permitted in thefe places to hawk about old clothes, or whatever they may have for fale.

CAMLEY, or KAMLEY; a garment in fhape like a carter's frock, made of the inteftines of marine animals, of linen, nankeen, or leather.

GORODNITSHIK; the mayor of a town.

GUBA; a bay.

KAMEN; a barren mountain; alfo a rock at fea.

KREPOST; in Ruffia, means a regular fort; but in Siberia, Kamtſhatka, and the iflands, it is ufed for any place walled in; and is a name frequently given to a place which was intended to have had a fortrefs; as Petro Pavloffky Krepoft, or the fort of St. Peter and St. Paul.

LAID, or LAIDENOI BEREG; a rocky fhore covered at high water.

MAMMONTS' TUSKS are found about the Siberian rivers and the fhores of the Icy Sea, and fcattered all over the arctic flats. They are full as large as thofe of the elephant, much more curved, and perhaps equal in quality. It appears that the animal is extinct.

MUYS,

Muys, or Mys ; a cape.

Noss ; a promontory.

Ostrog ; a fquare inclofure of palifadoes, about eight feet high ; replete with holes to point mufkets through : it generally has four entrances, with a tower upon each.

Ostrov ; an ifland.

Ozer, or Oser ; a lake.

Park ; a garment made like the camley, but only of the fkins of animals with the hair on, or with thofe of birds with the feathers.

Peredofshik ; a leader.

Polog ; a low tent ufed in a larger to fleep and fit in ; alfo a thin covering over a bed to keep away flies and mofquitoes.

Pood ; a Ruffian weight of forty pounds, equal to thirty-fix pounds Englifh.

Pristan ; a landing-place for goods.

Promyshlenik ; a hunter.

Quass ; a fermented liquor of plants, berries, roots, or meal, ufed as a drink.

Reka ; a river.

Retshka ; a rivulet.

Sazshen ; a fathom of fix feet.

Sheetiki, or Shitiki ; a large boat fheathed with plank, which is faftened to the timbers with twifted oziers ; the interftices are ftuffed with mofs, inftead of caulking ; and the feams are covered with laths of about two inches wide, to prevent the wafhing out of the mofs ; thefe are inclofed in the oziers. The name implies *fewn*, as they are made without nails or pegs.

Sloboda ; a large village with a church.

Sopka ; a peaked mountain.

Toion, or Toyon ; the Yakut name for chief, applied to the chiefs of all the heathen nations.

Ust, or Oost ; the difcharge of a river.

Utshenik ; a learner.

Verst ; a Ruffian mile, 104½ to a degree.

CON-

CONTENTS.

b   C H A P.

CHAP. VI.

CHAP. VII.

CHAP. VIII.

CHAP. IX.

CHAP. X.

CHAP. XI.

CHAP. XII.

b 2

Dress,

CHAP.

CHAP. XVI.

CHAP. XVII.

CHAP. XVIII.

CHAP. XIX.

CHAP. XX.

3

alarming

LIST

LIST OF ENGRAVINGS.

---

* The original from which this repreſentation was taken did not come within my own obſervation; it is, therefore, not explained in my narrative.

It is a piece of wood to which the claws of the Morzſh are faſtened; the hunters, covering themſelves with the ſkin of the head of the Morzſh, make a ſcratching noiſe on the ice with this inſtrument; the Morzſh approaches it, when the hunter takes his lance, and, throwing off the maſk, ſprings ſuddenly upon the Morzſh, and ſtabs it.

The

* The armour is made either of lathwood, with thin bone, or if they can obtain them, iron hoops in preference; they are faſtened together with the ſinews of ſeals, ſo that they will bend both ways, and are covered over with leather, which is bound on with thin ſlips of whalebone, which gives it the appearance of ſo many hoops. They are replete with loops and buttons, upon which they hang their bow, arrows, &c. ; the upper part occaſion-ally lets down.

ERRATA.

			for	read
Page 13	line	25	*for* foal	*read* fole
15	—	26	— 19th	— 9th
21	—	13 }	— Rhe	— Rheum
30	—	24 }		
26	—	22	— Pyat † Defetniks	— Piat-Defatniks †
42	—	16	— ripling	— rippling
—	laft line		— 800	— 300
45	line	11	— Iydomo	— Yudoma
—	—	24	— Ingigirka	— Indigirka
—	—	25	— irba's	— ifba's
—	—	23	— their cakes	— thin cakes
47	—	23	— Chap. V.	— Chap. VI.
54	—	1	— nodules	— needles
57	—	26	— faftened	— hung
59	—	25	— 29	— 19th
63	—	3	— Chap. VI.	— Chap. VII.
67	—	1	— Chap. VII.	— Chap. VIII.
82	—	1	— Chap. VII.	— Chap. IX.
99	—	1	— fouth-weft	— fouth-eaft
143	—	19	— north-eaft	— north-weft
144	—	9	— Treeh	— Trech
182	—	7	— Alcha	— Atcha
227	—	17	— Suchanin	— Luchanin
276	—	1	— Rakivinoi	— Rakovinoi
296	—	5		

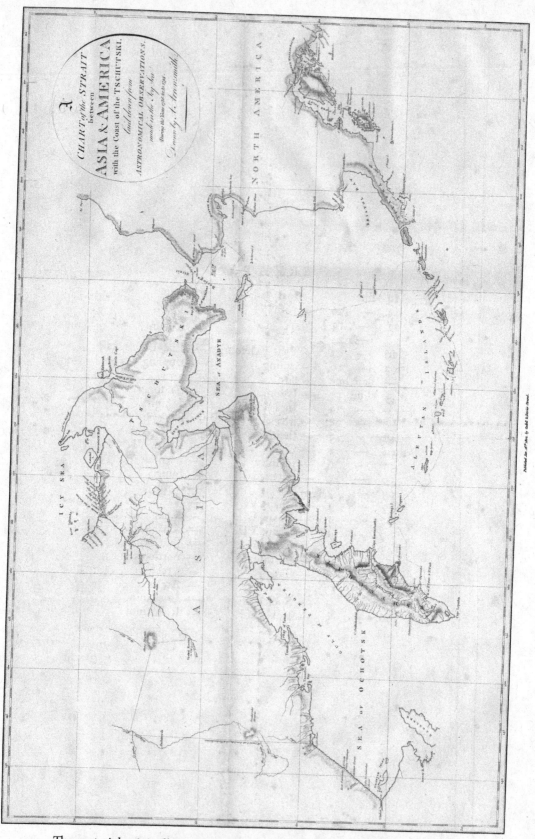

The material originally positioned here is too large for reproduction in this reissue. A PDF can be downloaded from the web address given on page iv of this book, by clicking on 'Resources Available'.

ACCOUNT

OF AN

EXPEDITION,

&c. &c. &c.

CHAP. I.

Departure from St. Petersburg.—Occurrences on the way to the City of Irkutsk.

In confequence of a mandate figned by the late Emprefs of Ruffia, Catharine II. directed to the College of Admiralty, and dated 8th Auguft 1785, appointing Captain-Lieutenant Jofeph Billings to the command of " A Secret Aftronomical and Geo- " graphical Expedition for navigating the Frozen Sea, defcribing " its Coafts, and afcertaining the Situation of the Iflands in the " Seas between the two Continents of Afia and America, &c. *" the Admiralty confirmed the officers chofen by the commander, and fupplied aftronomical and nautical inftruments, the charts and journals of all former navigators from the year 1724, and every other article confidered as neceffary.

* See the Introduction.

Early

Early in the month of September, Lieutenant Saretſheff was diſpatched direct to Ochotſk, with our ſhip-builder and his aſſiſt-ant, to ſelect and fell proper trees for conſtructing two ſhips, ac-cording to a plan of Mr. Lamb Yeames, ſhip-builder, in which he had conſidered the beſt means of accommodating the officers and crew. The injunctions laid on Captain Billings to explore the rivers and the inland country of Siberia, prevented our going by ſea from St. Peterſburg; beſides, the ſhips were to remain as tranſport veſſels, or armed cruiſers, in thoſe ſeas.

Lieutenant Saretſheff had orders to the Governor-General of Ir-kutſk and Kolivan, appriſing him of the purport of the expedition, and requiring his poſitive commands to the Governor of Ochotſk to ſupply men and neceſſaries to forward the buſineſs.

The whole party was ſent off in ſmall detachments by the middle of October; on the 25th day of the ſame month, 1785, I departed from St. Peterſburg, with Captain Billings and our ſur-geon, Mr. Robeck, at eight o'clock in the evening; in very rainy and windy weather; which made the roads ſo indifferent, that we did not reach Moſco till the 5th of November at eight o'clock in the morning. I forbear to make any remarks reſpecting the road, the villages, &c. as every circumſtance is well deſcribed by Mr. Coxe; my intention is, to be more particular when I arrive at places leſs known.

In this city Lieutenant Hall, the ſecond in command, was or-dered to wait the arrival of Captain Billings; the other parties proceeded by Kazan to Irkutſk.

We

We purchaſed a great number of articles neceſſary for our travels; received medicines for the uſe of the expedition, from the
Government General Repoſitory of Drugs; and forwarded our baggage to Kazan by Lieutenant Hall, who left Moſco the 10th of December; and on the 15th, at four o'clock in the afternoon, Captain
Billings proceeded with our head ſurgeon, (Mr. Robeck,) Sturman
Batakoff, and myſelf, attended by ſoldiers in carriages and on
ſledges.

The road was barely covered with ſnow; and on the 18th we
arrived at Paulova, containing about 2500 houſes, ſome few very
elegantly built of brick, and five churches. This may be conſidered as the Birmingham of Ruſſia, and is, with all its inhabitants, the private property of Count Sheremetoff. The people
are all manufacturers of hardware and traders, have an immenſe
number of well built veſſels, and carry on a very extenſive trade
in the Caſpian Sea. This (Slobôda) large village is ſituated on
the river Oka, near its diſcharge into the Volga. We here purchaſed knives, ſciſſars, buttons, &c. as preſents to the natives of
ſuch places as we might touch at in our voyage.

At the diſtance of 625 verſts from Moſco we entered an oak
wood, chiefly of middle-ſized trees, near the Tſheremeſe village
of Scartog, travelled 75 verſts through it, and arrived on Monday
the 22d of December, about eight in the evening, at Kazan, where
we found our whole party in good health and ſpirits.

Kazan is a regular and well-built city on the river of the ſame
name, three verſts from its diſcharge into the Volga, and ſituated
in latitude 55° 43', and longitude 49° 15' E. from Greenwich.
The inhabitants, who are chiefly merchants, conſiſt of Ruſſians,

Tartars,

Tartars, and Armenians, and carry on a very confiderable trade.

Numbers of noblemen refide here in the vicinity of their eftates; and others, who find motives for retiring from the capital, alfo choofe this city for their place of refidence. We obferved that the greateft harmony reigned among them, with unbounded hofpitality, efpecially to us as travellers. It becomes us particularly to acknowledge the great kindnefs that we experienced from the Prefident of the Admiralty and Director of the Dock Yard, Admiral Zfhemtfhuzfhnikoff, whofe houfe we made our head quarters and general rendezvous. This nobleman was in England about the year 1770, engaging tranfports for the Ruffian fleet under the command of Count Orloff, and was very much attached to the Englifh. At his friendly manfion we ufhered in the year 1786, and in our company he made a point of drinking the health of the King of Great Britain, and fuccefs to his fleets. The recollection of kindneffes that he received in England gave him enthufiaftic pleafure, fomething, I believe, like what I feel at this inftant on recollecting the favours that I received from him. His table was always profufely covered, and his wines were of the beft flavour.

On the 6th of January, after attending the ceremonies of the day (as defcribed by Mr. Coxe), we dined with the Governor. The ftrange mixture of his company I think worthy of notice. The bifhop of Kazan (a very learned divine, a great friend to the poor inhabitants, and the founder of a fchool for their children), the Mahometan Chief Prieft, a German Lutheran Prieft, with feveral natives of Ruffia, England, France, and Italy; and, though the good humour of the company was not increafed by

the

the luxury of the table, or the excellence of the wines, it did not suffer by any reflections on our host's want of generosity.

All the neceffaries and fome of the luxuries of life are in great plenty in this city, and at a very moderate price.

The command was difpatched from hence as follows:

1ft Party, 31ft December, 6 Kibitki fledges.
2d ——— 2d January, 6 Do.
3d ——— 4th Do. 6 Do.
4th ——— 7th Do. 6 Do.

with orders to make the beft of their way to Irkutfk.

On the 9th of January Captain Billings reported to the Admiralty the ftate of his command, and requefted a frefh fupply of barometers, every one that we had with us being broken, owing to the bad ftate of the roads. One of our medicine chefts alfo broke through the ice in croffing the Volga, which fpoiled a great part of the contents; and thefe, of courfe, our furgeon required to be replaced and fent to Irkutfk.

On Saturday the 10th we left Kazan: the roads were now good, and well covered with fnow. At the diftance of 18 verfts we entered a wood of very fine oak *, through which we travelled 34 verfts. The face of the country continued hilly and moderately wooded with fir, common pine, and birch. The in-

* On my return by this road, in January 1794, I was furprifed at feeing the country cleared of every tree, and lying wafte; not even a bufh being to be feen; which was pretty much the cafe with the wood near Scartog on the other fide of Kazan.

habitants

habitants are Ruffians, Tartars, and Votiaki. The Tartars are Mahometans, and very clean in their perfons and habitations. The women are, in general, very handfome, and drefs extremely neat. They are induftrious, honeft, and peaceable ; and, under their management, a piece of ground of a given extent will pro- duce nearly twice as much as the Ruffians obtain from an equal quantity. All the villages are built in vallies on the borders of rivers, furrounded with gardens and cultivated fields.

On Monday the 15th we reached Kungur, a city containing 1800 wooden houfes on the eaft fide of the river Tulva, latitude 57° 20′, longitude 56° 50′ E. 2160 verfts from St. Peterfburg: and here the Virchoturien mountains commence.

Atchinfky Krepoft, on the confines of Siberia, is 88 verfts be- yond Kungur; and in this neighbourhood are the iron works of the different rich proprietors living in St. Peterfburg, &c. The inhabitants appear particularly ftrong and healthy; their houfes are very clean; and I faw feveral men who were not very infirm at between 80 and 90 years of age. The woods that we had hitherto paffed confifted of fir, common pine, poplar, afp, and birch : here we obferved alfo the larch pine.

On the 17th January we arrived at the famous city of Ekate- rineburg, on the Uralian chain, through which the river Ifet flows, and works the gun, anchor, and iron foundries, faw and coining mills, and lapidary, &c. belonging to Government. This city ftands in latitude 56° 50′, longitude 60° 17′ 10″, and con- tains about 2000 houfes, fome very elegantly built of ftone, five churches, fchools, &c. Provifions are here extremely cheap; fifh, fturgeon, beluga, and large quabs (nalime) 20 copeaks the pood, beef
50 cop.

50 cop. rye flour 32 cop. * The laft article, they faid, was very dear, having had but a fcanty growth for the laft three years. The poorer forts, and convicts, of which only a few work at the mines, find a plenteous and cheap fupply of the falted omul, which appears to me to be a fpecies of herring, but twice as large as the ordinary fort. The circumjacent mountains afford much interefting entertainment to a naturalift, whofe refearches are fre- quently rewarded with new difcoveries of variations in the works of nature. Exclufive of minerals and malachites (the largeft ever heard of was found here, weighing 107 poods, or 3852 pounds· weight,—Pallas), here may be feen an aftonifhing variety of white rock cryftal, with capillary fhörl of different colours; that with the red was called by Pallas the hair of Venus; the green was named by Guthrie Thetis's; the flaxen, Cupid's; the black, Proferpine's; and a dark topaz, containing white fhörl, Saturn's hair; ame- thyft, topaz, the Siberian diamond, jafper ftriped and plain, por- phyrys, &c.

We now found the face of the country more level, and the woods very much on the decreafe, and pretty well inhabited by Tartars and Ruffians; the former of whom, befides cultivating the earth, make and fell very neat worfted carpets.

On Thurfday the 22d we arrived at Tobolfk †, containing 2300 houfes and 23 churches, chiefly of wood, latitude 58° 12′ 20″, longitude 67° 19′ 10″, oppofite the difcharge of the Tobol

* A copeak may be reckoned about a farthing; a pood is 36 pounds.

† This city was reduced to afhes in the year 1787; but on my return (in 1794) it was rebuilt on a regular plan; the ftreets wide, with churches, and a great number of houfes, of brick. Its fituation is low, backed by a rifing ground which projects over the Irtifh; and the fortrefs is built on its point.

into

into the Irtifh. It is a place of very confiderable trade, frequented by Samoyeds, Tartars, Ruffians, and Siberians; and provifions here are rather cheaper than at Ekaterineburg.

I obferved that we had now croffed the whole of the Uralian chain of mountains.

Bad weather detained us here three days, during which fhort ftay we experienced the hofpitality of the Governor General Kafhkeen.

On Sunday the 25th we left Tobolfk, and found the country low, marfhy, and woody, but well inhabited as far as the Defert of Baraba (Barabinfky Step), fituated 3512 verfts from St. Peterfburg, and 570 verfts in extent. This was not inhabited, but the Emprefs about ten years ago built villages all over it at the diftance of 20 to 25 verfts. The whole extent almoft is low and boggy; fome elevated fpots, however, produce ftunted birch, brufhwood, and a little grafs. The weftern half is well watered by the lake Kamyfhova, the rivers Om, Tartas, and Kain. The latter is about the midway; and here the town of Kainfk is built for the protection of travellers. It contains 125 houfes and a church, all of wood. The inhabitants of all the villages are convicts. The eaftern half of the defert is almoft deftitute of wood; nor is there any water, except in ponds, fetid even at this feafon. Wells are dug, but thefe prove falt and bitter. The people, therefore, melt fnow in winter, and collect rain in fummer. The moft extreme poverty, dirt, and mifery prevail over the whole defert, which is terminated by a rifing ground, where Nature fuddenly affumes the moft luxuriant change, prefenting a woody country, replete with meadows, corn fields, and well

built

built villages, inhabited by clean, healthy, and affluent Siberians.

On the 4th of February we reached Tomſk, a city containing about 1500 houſes (latitude 56° 29', longitude 85° 3'), on the river Tom. Here we ſaw Mr. Patrin, the gentleman appointed by her Imperial Majeſty, on the recommendation of the very learned Dr. Pallas, to accompany our travels as naturaliſt and botaniſt. He was on his return to St. Peterſburg, with a view of retiring to France on account of his health, the ſtate of which rendered his longer reſidence in theſe parts dangerous, and bereaved us of an invaluable companion.

The inhabitants of theſe parts are farmers, graziers, and carriers; and have a famous breed of horned cattle, with which, as well as with butter, they ſupply all the northern and eaſtern diſtricts of the empire. They are wealthy, hoſpitable, healthy, and clean, and live under no controul of individuals, only paying a trifling ſum to the Captain of the diſtrict, for Government. The Siberians throughout are more induſtrious and independent than any Ruſſian peaſants, live more comfortably, and drink home-brewed beer in addition to quaſs. The women are remarkably clean; and I never entered any houſe in travelling, night or day, but I found them ſpinning flax. I have frequently aſked them why they worked all night; and always received one general anſwer, " that the days were ſhort." Inſtead of candles, they burn laths of birch wood, which they call lutſhinka; a portable ſtick, about five feet high, with a foot to it, ſerves as their candleſtick; three nails are driven into the top, forming a triangle; the lath is ſtuck between the nails obliquely, and burns about four or five minutes: but when they have gueſts they burn a candle, the nails ſerving as a ſocket.

C

Their

Their neighbours, to the fouth-eaft, are fome tribes of Mongals extending to China; and a private trade is carried on, through this channel, by a few of the more intelligent Siberians.

On the 14th February we arrived at Irkutfk, in a froft of 18° of Reaumur, and found all the Command fafe. The Governor General Jacobi being abfent, a courier was difpatched to him at Barnaul, to inform him of our arrival.

On the 15th, in the morning, the thermometer indicated 28° below o of Reaumur for about two hours, when it rofe to 20° and 18°: 30° is the greateft extremity of cold ever remembered here.

CHAP.

CHAP. II.

A succinct Account of our Transactions at the City of Irkutsk—State of our Command—Additional Hands taken there.

WE were now arrived in the capital of Siberia, and entering on the first scene of real business, of a nature more extensive than this city ever before experienced. Every necessary article for constructing two ships of 85 feet keel was to be received here, except timber: iron, cordage, ammunition, provision, liquor, stores; clothing from head to foot, on a moderate computation, for five years; candles, soap, and every the most trifling commodity necessary for each individual officer, over and above the ordinary allowance for a Command of three hundred men, to be transported by water and land upwards of four thousand versts. It was likewise necessary to stow every article in the most secure method in packages, weighing only $2\frac{1}{4}$ poods each, or 90 pounds English weight, to facilitate their transport on pack-horses from the city of Yakutsk, both to Ochotsk and to the river Kovima, where we had to build a vessel of 50 feet keel, with boats and baidars, for navigating the Icy Sea.

No time was lost in ordering the instruments in very great abundance, with an extra number to serve as presents to savages. Some hands were sent, under the command of an officer, to build vessels at Katshuga Pristan, on the river Lena, to transport the command to Yakutsk, &c.

On

On the 26th of February we began to receive hatchets, hammers, and every other inftrument ufed by artificers in a dock-yard.

At half paft three o'clock in the morning of the 3d of March, the city experienced a fmart fhock of an earthquake, which lafted about three feconds, but was not attended with any bad confequences. Earthquakes here are frequent, but not violent.

On the 5th the courier returned from the Governor General, with orders for the Governor to comply with all the demands of the expedition. To forward the bufinefs, Captain Billings increafed his Command, agreeable to his inftructions. From St. Peterfburg it confifted of,

Captain Jofeph Billings, Commander ;
——— Robert Hall ;
——— Gabriel Santfheff ;
——— Chriftian Bering ;
Skipper Affanaffy Bakoff, to rig the veffels, and take charge of all ftores ;
Steerman Anton Batahoff, }
——— Sergey Bronnikoff, } Mafters ;
Surgeon Mich. Robeck ;
——— Peter Allegretti ;
Drawing-Mafter Luke Varonin ;
A mechanic ;
Two under fhipbuilders ;
Two furgeons' mates ;
One mafter's mate ;
One boatfwain ;

Three

Three Court Yagers, for ftuffing birds and beafts;

Eight petty officers;

Seven foldiers, Streltfi;

And myfelf, as private fecretary, and journalift:

In all thirty-fix.

The following hands we engaged here:

Vaffiley Diakonoff, } for keeping accounts, and tranfacting the
Fedor Karpoff,   }   writing bufinefs for the command, in
                    }   Rufs;

Lieutenant Polofoff, of the army; [He had occupied a place of truft in Igiga, and was recommended as a ufeful hand among the Tfhutfki.]

Six petty officers from the Navigation School at Irkutfk;

Three men for conftructing leather boats, or baidars, for river navigation;

One turner;

One lockfmith;

Fifty Coffacs, and one Sotnik;

Two drummers:

In all, fixty-nine, in addition to the St. Peterfburg Command.

March 7.—The perfons compofing the Command were advanced a rank, agreeable to the Imperial mandate.

All hands were now employed in packing up inftruments, &c. in boxes, which were afterwards covered with canvas, pitched all over, and fewn up in foal leather, to prevent water from fpoiling the contents in time of rain, and in fording of rivers. The leather was ultimately defigned for fhoes and boots.

March

March 16.—The ice of the river Angara broke up, and we had fine and mild weather.

April 16.—The weather being favourable, Captain Billings erected his aftronomical tent at the fouth-eaft extremity of the city ; and, by many fets of obfervation of the fun's and moon's diftance, his longitude proved 103° 46′ 45″ eaft of Greenwich, latitude 52° 16′ 30″.

Irkutfk contains 2500 houfes, chiefly of wood, 12 ftone churches, a cathedral, and two monafteries ; befide which, there are feveral public buildings, an hofpital, an inoculating houfe, a feminary for the ftudy of divinity, a public fchool, a library and collection of curiofities ; alfo a theatre, of which the performers are all young men and women natives of Irkutfk. The reprefentations are chiefly confined to national pieces, which they get up with aftonifhing propriety ; and they have very excellent muficians belonging to the different regiments, befides the band of the Governor General.

This city ftands on a low fpot of ground, oppofite the difcharge of the Irkut into the Angara. The latter river bounds it to the weft ; the Ooſhakofka to the eaft and north-eaft ; and to the fouth, high, pleafant, and fertile land.

The ftreets are ftraight and uniform. The fhops are in the heart of the city, an elegant fquare pile of brick building, under piazzas which fupport warehoufes. The butchers' fhops are in the eaftern extremity,. where the flaughter-houfes are built over the Ooſhakofka. Near this is the fifh-market ; alfo a bazar, or renok, for vegetables, corn, bread, butter, flour, pedlary, and wooden ware.

9

The

The latter is a place of refort of the Burati, who hawk about fables, martins, otter, and other furs. I calculate the number of inhabitants at about 20,000. The merchants are numerous and affluent; and a confiderable trade is carried on with the Chinefe, which is fo well defcribed by Mr. Coxe, that all I can add to his remarks on the fubject is, that the prices of articles are now about three times as high as when he mentioned them. Here the affortments of furs are made, which are brought from America, and the northern parts of the empire, in the following manner: The inferior and worft coloured fables, the fox fkins, from the Aleutan iflands; the fecond fort of fea otter, river otter, &c. are allotted to China *. Such as are defective and weak in the hair, as alfo inferior forts, are fent to the fair at Irbit; and the very beft are fent to Mofco and Makaria, where they meet with ready purchafers among the Armenians and Greeks.

The works belonging to this place are, a glafs-houfe under the particular infpection of the learned Profeffor Laxman, fituated near the Baikal lake; a diftillery, about 60 verfts north of the city, in which the annual average quantity of brandy made is 60,000 ankers; falt works at three fprings, which fupply the neighbouring country; a cloth manufactory, where eleven looms were in conftant employ, but now only one is at work for very coarfe cloth: this is the property of the merchant Siberakoff; the others belong to government.

On the 28th April, we began fending off the Command and articles to Katfhuga Priftan; and by the 19th May all were difpatched from hence.

* The Chinefe colour fables, and other furs, fo artfully, that the deception is not obfervable: in confequence, they will only pay a low price, and always give articles of an inferior quality the preference.

In. me-

Immediately on our arrival at Irkutſk, Profeſſor Laxman of-fered to accompany the expedition as naturaliſt and botaniſt : Captain Billings, however, did not accept the offer ; and it was only the day before his departure that he ſent me to Dr. Merck (belonging to the hoſpital here) to aſk if he would go with us in that capacity, which he immediately agreed to, but confeſſed that he was not a proficient. Mr. John Main, an Engliſhman (alſo a medical gentleman), volunteered to go as his aſſiſtant. Thus were matters ſettled ; and all the neceſſary articles and books given by Profeſſor Pallas for Mr. Patrin, were now ordered to the diſpoſal of Dr. Merck, who left Irkutſk with us the very next day.

Before I quit this place, which, according to the information of my acquaintance, is the laſt of any conſequence that I am likely to behold till my return, I ſhall attempt a ſummary ſketch of its inhabitants, and their mode of living.

Lieutenant General Ivan Varfolomitſh Jacobi, Governor Ge-neral, poſſeſſing the powers of Viceroy ;

Major General Lamb, his Aſſiſtant Governor ; and

Mr. Medvedeff, a very rich individual, keep open houſe, and give a dinner and ball each once a week. The remaining days are paſſed in viſiting other opulent inhabitants, either in conſe-quence of invitations, or in the way of friendly call. The ſet is never broken, though ſometimes divided into ſeveral branches ; but they are always united at every invitation. At dinner a band of muſic induces an harmonious circulation of the glaſs.

The ſociety eſtabliſhed, and the liberal hoſpitality of the firſt order of inhabitants, is ſuperior to that in any part of Ruſſia, and

really

really feems to infufe a fpirit of confequence into the minds of the lower fort of people. I think that their fchools and theatre contribute much to this; but moft of all the tutors to the children of the more opulent. Thefe generally confift of Poles, Swedes, French, and fome of the Jefuitic order, who have been under the neceffity of travelling.

Numbers of mechanics, artifts, and artificers of great abilities, whofe exertions were felfifh in Ruffia, here exert themfelves for the benefit of the community; and, as merit is the chief introduction to independent fociety, fo all who poffefs it meet with liberal encouragement; and, unlefs their characters are fullied by acts of criminality, they are countenanced and fupported. The unfortunate are generoufly diftinguifhed from the villainous.

The officers here, both military and civil, are very numerous; the former, in confequence of this being the feat of government in the vicinity of the Chinefe and Mongal territories; the latter, on account of the numerous courts of juftice, and the neceffary diftributions to be made for the vaft extent of its jurifdiction. I fhall rate thefe in two claffes; for rank is only a fecondary recommendation here: the gentleman, who behaves himfelf with propriety, though poor, is completely independent, and every houfe is open to him; while the worthlefs are only attended to in the execution of their duty, and then with great referve.

In this town there are neither inns nor coffee-houfes; but no ftranger, who behaves himfelf with common civility, will ever be at a lofs for a home. I had very good quarters allotted me by Government, in which I had only refided a few days, when Brigadier General Troepolfky invited me to accept of apartments and

D attendants

attendants in his houfe : his lady repeated the invitation, which I begged they would allow me to refufe. They then fent me every neceffary to my lodging, which really compelled me to accept their firft offer, to fave them greater trouble. Their manfion was ever after my home ; and their friendfhip will always remain indelibly impreffed on my mind. All kinds of food are cheap, as are fpirituous liquors and home-brewed beer. Wines are dear. Many luxuries are imported from China ; and filks, cottons, linens, furs, nay Englifh cloths, are moderate.

Throughout the whole of Siberia, hofpitality prevails in the extreme. A traveller is perfectly fecure on the road, and certain of a hearty welcome wherever he puts up, let the cot be ever fo homely. But whether this hofpitality will continue when they arrive at a certain ftate of refinement, to which they feem advancing with incredible hafte, remains for future times to difcover ; as alfo whether the expanfion of ideas may not lead to the extenfion of territory, and other formal eftablifhments.

In the morning of the 10th of May we had a heavy fall of fnow, which lay on the ground two hours. By noon it cleared up ; and in the afternoon, at fix o'clock, Captain Billings, Dr. Merck, and myfelf, left Irkutfk, accompanied by Count Manteuffel, Mr. Haak, &c. The Governor General had prepared a farewel fupper for us at his villa, 18 verfts from the city, where we paffed the night. The next morning, at fix o'clock, we took leave of our friends, with a moft grateful fenfe of the multiplicity of favours that we had received during a ftay of three months. Our road lay acrofs the Buratfkoi Step, fo called from the immenfe herds of cattle and horfes that the Burati graze here. Thefe are of the Balagan tribe, and, with the Charintfi,

feem

feem outcafts from the race of Mongals ; for they have no lamas, nor letters, but are complete demonolatrians, and confult their forcerers: all other tribes of Burati are intelligent people, have lamas, or priefts, and are ftrict obfervers of their religious rites and ceremonies ; their worfhip being performed in the Tungut-fki language, in which all their laws are written. They have different places for particular rites. The ceremony of an oath, or rather a curfe, to prove their innocence, if fufpected of a crime, is performed on fome felected mountain : formerly, a very remarkable one fituated near Kiachta, and called Burgutta, was their favourite place ; but by a late treaty this mountain fell to the lot of the Chinefe, which caufed great diffatisfaction to the Burati, and was followed by defertions.

We arrived on the 12th, at two in the afternoon, at Katfhuga Priftan, a village containing only 15 houfes, fituated on the river Lena, in latitude 53° 26', longitude 107° 2', 230 verfts from Irkutfk. Here we found nine barks of from 10 to 15 tons burthen, ready to convey us and our baggage to Yakutfk. The number of packages amounted to 2600, exclufive of fail cloth, cordage, &c. not yet fent from Irkutfk ; to forward which we left behind us Sturman Bronnikoff. Our guns, anchors, &c. were not yet arrived, nor did we expect them till the next year.

On the 14th May, in the evening, we had loaded all our goods, and got every thing ready for our departure down the river.

CHAP.

CHAP. III.

Departure from Irkutſk.—An Accident.—Remarkable Cave.—The River Lena and its Tributaries.—Arrival at Yakutſk.

ON Friday the 15th of May, at eight in the morning, Captain Lieutenant Hall ſet ſail with three barks. At five in the afternoon, Captain Billings followed with the reſt, giving the charge of them to Captain Lieutenant Bering. He then embarked in a dosſhennik * with Dr. Merck and myſelf, Count Manteuffel ſtill accompanying us. The night was very dark, with thunder, lightning, and heavy rain. At midnight we arrived at Vircholenſk, (an oſtrog containing 100 houſes, and two wooden churches, famous in theſe parts for its manufactory of coarſe worſted ſtockings and night-caps) 35 verſts from the place of our embarkment, and made faſt to the ſhore. Captain Bering, a-head of us, ran foul of a merchant's bark, and ſunk in nine foot water cloſe to ſhore. We employed all hands in unloading the cargo, and found that, notwithſtanding every precaution uſed, the boxes were not water proof. The moment the accident happened, the pilot leaped overboard, and ſwam to ſhore.

At ſeven in the morning of the 16th, Count Manteuffel went, with Dr. Merck, to ſee a remarkable cave, 15 verſts from hence.

---

* A veſſel calculated for accommodating a family, with baggage, down the Lena. It contains three cabins : one at the forecaſtle, with a ſeparate entrance ; one in the midſhips, and the other aſtern. It is built of boards without a keel, flat bottomed, about 35 to 40 feet long ; rows and ſteers with long ſweeps, two men to each ; is furniſhed with a maſt, and one ſquare ſail, and named from *doſok*, a board.

They

They returned at two in the afternoon, and gave the following account of their expedition.

The mountain is called Khacharchai by the Burati, and the cave is fituated about the middle of the afcent, furrounded by large trees of pines and birch. It is about one yard high, and half a yard wide; and the outward borders, as alfo the infide, as far as the eye can reach, are covered with a thick coat of ice. A thermometer in the fhade was 14° above the freezing point; while another, placed about a fathom within the cave, ftood 4° below it. A pretty frefh wind iffues from the cave, which, according to the account of the guide, freezes all the fummer, but thaws in winter, when a warm vapour fupplants the chilling breeze. Among a great variety of plants near the place, the Rhe Rhaponticum is the moft prevalent. About half-paft three this afternoon we took leave of Count Manteuffel, who returned to Irkutfk.

Mr. Bering's bark being repaired, and reloaded from two other barks, which were left under the care of Mr. Main to wait the drying of the damaged articles, Captain Billings immediately pro-ceeded with the other four.

I fhall not weary the reader's patience with a circumftantial account of every day's progrefs; as I mean to ftate, in an Appen-dix, every ftage that we paffed from Mofco, noting the number of houfes, and of verfts, as alfo the time of our arrival and de-parture.

Continuing the plan that I at firft adopted, I fhall give a very brief account of this river, to avoid fatiguing my readers, and to

encourage

encourage them to follow me through all the mazes of my progress.

The Lena takes its rise from an inconsiderable lake situated between the mountains near the Baikal, about 100 miles west south-west of Katshuga Pristan (wharf). It flows in a gentle and uninterrupted stream, though here and there impeded by shallows at a late season, to about the distance of 300 miles from its source, when it deepens considerably. The direction is very winding, but pretty uniformly east north-east to Yakutsk, and nearly north from thence to its discharge into the Icy Sea, about the latitude 71° 30', and longitude 127° east of Greenwich, after a course of 3450 geographical miles. The appearance that it assumes is continually varying; in some places mountains bound the channel on both sides, clothed to the summit with stately pines; in others, they are barren, projecting into the river, and turning its course; taking fantastic shapes, resembling ruins of large buildings, towers, and churches; the chasms overgrown with hawthorn, currant-bushes, dog-roses, &c. In some places the mountains retreat inland for miles, forming a back ground to extensive plains, and exposing a miserably built town, surrounded with cornfields, gardens, and pasture grounds, with a few herds of cattle grazing; these openings are frequent at unequal distances of 5 to 40 versts from each other, and are always occupied by villages as far as Olekma, 1800 versts from Katshuga: all beyond is desolate, except a few huts inhabited by convicts, who have the charge of horses for the post, and the towns of Pokroffsky, Yakutsk, and Giganfk. The best of them is only a collection of huts inhabited by priests and their attendants, officers and Cossacks, who teach obedience, and enforce the payment of tribute from the wander-

I

ing

ing tribes of Tartars that infeft the neighbourhood, and of whom I fhall have occafion to fpeak hereafter.

The rivers that flow into the Lena are,

The Ilga,   -   170 verfts from Katfhuga.

Koot,   -   469——Very near the eftuary of this river is a falt lake, which is very fhallow; and works, the property of the prefent Ifpravink of the diftrict, at which one boiling produces 1080 pounds weight of falt.

Marakofka,   -   601.

Makarova,   -   690.

Kiringa,   -   778.

Vitima,   -   1178.——This river flows from a lake eaft of the Baikal. It is nearly equal to the Lena, in width, depth, and extent; and is famous for fables, lynx, fox, ermine, fquirrel, and deer.

The fables of this river, and of the Momo, which falls into it 300 verfts from the difcharge, are very valuable, and of a fuperior quality. Numbers of Tungoofe travel about here on the chace. Three verfts up this river are the mountains that produce talk. I have feen fpecimens formerly found, 28 inches fquare, and tranfparent as glafs: what is now found

found is very ſmall, but perfectly pellucid. All the windows of theſe parts
are glazed with it.

Pellidui,   -   1202——Alſo famous for the above mentioned animals, and the laſt place that
produces corn.—*N. B.* Sparrows and
magpies were not ſeen further north.
They only came here about five years
ago, after the ground had begun to
be cultivated.

Nuye,   -   1475.

Yerba,   -   1505.——Here we ſaw the firſt Yakut or
Sochalar. He was our pilot ; very
communicative, and good-natured.
In this neighbourhood the river is
replete with iſlands, having on them
temporary Tungooſe habitations for
fiſhing.

Patama,   -   1575.
Oonaghtak,   -   1595.
Olekma,   -   1822. ⎤ I ſhall refer to theſe rivers, and deſ-
Aldan,   -   2600. ⎬ cribe them on a future occaſion, in
Viluye,   ⎦ an account of the Amur.
Beſides ſeveral rivers farther north, of no material conſequence.

Our navigation only extended to the town of Yakutſk, 2390
verſts from Katſhuga. We arrived in this town the 29th May, at
ſeven P. M., and immediately ordered the loaded veſſels to croſs
the river to the plains ; whence the goods are to be forwarded on
pack-

pack-horfes. The next day the other barks arrived fafe at the above-mentioned place.

To tranfport the articles that we had with us, and the parties, acrofs the uninhabited country to Ochotfk, and to the river Kovima, two thoufand horfes were demanded of the commandant, or military go-vernor, whofe bufinefs it is to enforce the fame by a paper figned by himfelf to the court of the interior, or land diftrict. The Go-vernor General of Irkutfk, Jakobi, had clofed a contract with the merchant Siberakoff for provifions, confifting of flour, peafe, grits, oatmeal, meat, butter, falt, vinegar, brandy, &c. an 18 months fupply for 100 men ; and the contractor bound himfelf to deliver the fame at Virchnoi Kovima, by the firft day of Auguft next. Sail-cloth and raw-hides, for making baidars *, were alfo to be delivered by him, loads for more than 2000 horfes.

The fame gentleman clofed the contract for the delivery of pro-vifions, &c. at Ochotfk, a three years fupply for 250 men ; tallow for candles, greafe, pitch, tar, &c. &c. Befides the immenfe number of horfes wanted to tranfport the articles mentioned, our guns, anchors, cables, cordage, fail-cloth, cloth, and flops, with ammunition of all kinds, ftill remained to be forwarded from Irkutfk.

Thefe horfes were to be obtained from the Yakuti inhabiting the diftrict of Yakutfk, and the Viluye. I think it neceffary to re-mark, that to every three loaded horfes, a fpare one is allowed ; and a guide has charge of only fix under loads, two fpare ones,

* Baidars are boats very flat bottomed, the frames made of twigs, and covered with leather : they are fo light, that four men can carry them with eafe, and are rowed with fix or eight oars.

E   and

and one upon which he rides; fo that where 2000 horfes are demanded, 3000 are employed. The leffer number is paid for at the rate of one copeak a verft *per* horfe: the average rate of travelling is 20 verfts each day.

I was furprifed at the aftonifhing activity that appeared in every officer civil and military, from the firft to the laft, in volunteering to go to fome tribe or other of the Yakuti to procure horfes; but the grand point could not long remain a fecret; for their excefs of zeal (as is the cafe in moft zealous meafures) led to an open breach between the military and civil government, which lafted juft long enough for each to explain the interefted views of the other. After they had done this in the moft forcible manner, and completely vented their rage, they began to deliberate; when it appeared, that both parties agreed upon the whole, and that the extent of the commiffion would admit of a general divifion: fo they foon made up their minds to fend fome of each party with official authority. Univerfal harmony was immediately reftored; and each commiffioner went with a full refolution to ferve his employer in the beft manner he could, with propriety to himfelf.

This was the firft town in which I obferved the officers from the higheft to the loweft ranks form the poorer fet of inhabitants; while the Coffack Sotniks *, and Pyat † Defetniks (petty officers), were the moft affluent. They are chiefly Sinboyarfki ‡, well acquainted with the languages of the Yakuti, Tungoofe, &c. and are always employed in offices of truft. We found the beft entertainment among them; at the fame time they appeared the

---

* Commanders of 100.   † Commanders of 50.

‡ The loweft clafs of gentility, conferred upon the Siberians for fome particular achievement or difcovery.

more welcome guefts at the habitations of their fuperior officers, where they pay their refpects in the moft flattering manner, and never experience a refufal in a proper application for an advantageous miffion. I had my quarters at the houfe of the commandant, with Captain Billings.

In the evening of the 7th June, the firft party was difpatched, under the command of Mr. Bakoff, with 136 loaded horfes. Captain Lieutenant Hall had the charge of the parties to Ochotfk, and Captain Lieutenant Bering of thofe to the Kovima.

Yakutfk contains 362 wooden houfes, five churches, and a cathedral. A monaftery is now building, latitude 62° 1′ 50″ north, longitude 129° 34′ eaft, on a fhallow branch of the river Lena, three verfts weft from the main channel (*which is four verfts acrofs*), fituated on a low fandy plain, fixty verfts long, in a direction north-weft, and eleven broad, producing chiefly wormwood, thiftles, a few flowers, and wild onions; here and there clufters of hawthorn bufhes and oziers, with currants, dog-rofes, and rafpberries. It is bounded to the weft by a ridge of inconfiderable but woody mountains, from which the inhabitants obtain firewood.

Never was there a town in a worfe fituation than this. The branch of the river on which it is built is dry by the middle of July, and continues fo all winter, the inhabitants having to go the diftance of three verfts for water. Although the river abounds with fifh, they receive their fupplies of that article, as alfo of meat, from about the Viluye, 400 verfts down the river. Vegetables are brought them from the neighbourhood of Kiringa, 1650 verfts up the river.

E 2

In

In the month of June every neceſſary of life is brought hither down the Lena; and this is called the Yarmank. During this time every trader has permiſſion to hire a public ſhop, and ſell his ware; and this is the time at which the opulent lay in a twelve months' ſtock; for at the expiration of the month, the privilege of trading is only veſted in the hands of the burghers, who make their own prices: theſe conſiſt of five or ſix. Siberakoff, the contractor, has a houſe here, and at preſent occupies it himſelf, to ſuperintend in perſon the diſpatch of the articles for the expedition. During our ſtay, he may be ſaid to have kept open houſe for the entertainment of the chief inhabitants and our gentlemen.

On Tueſday the 9th June we took leave of our friends in this city, after a ſtay of 10 days, and croſſed the Lena, which is four verſts over, to the plains, called the Yarmank, from being the general rendezvous for all travellers, traders, and tranſport goods, to the eaſtern and north-eaſt parts of the empire. Here are extenſive meadows, producing graſs in abundance. The moſt prevailing plant that I obſerved was wild flax, ſome with white, and ſome with blue flowers; and a remarkable plant which the Ruſſians call Zemlennoi Laudon, or frankincenſe of the earth: this is not a gum, but an aromatic root, given to children and to adults for pains in the bowels; its ſmell is very like that of ſnakeroot, though in appearance it is not ſo fibrous. Maiden-hair grows in abundance, and is collected, dried, and uſed by the Coſſacks inſtead of hops. The Yakuti alſo make uſe of it occaſionally, with juice of berries and water, to drink. Some ſandy places are overgrown with horſe-radiſh and wild onions.

On Wedneſday the 10th June, at nine o'clock in the morning, all our baggage and food for the road were loaded on pack-horſes,

about

about five poods on each, and we commenced an equeftrian route. Our party confifted of Captain Billings, Dr. Merck, Mr. Robeck, Mr. Main, and myfelf, two petty officers, and nine privates, with an adequate number of Yakuti to take care of the horfes and ferve as guides. We travelled this day 28 verfts, to the folitary habitation of a Yakut, where we halted and pitched our tents for the night. The next day we made 49 verfts.

On Thurfday the 16th of June we arrived, about eight P. M. at the river Amga, or Anga, which falls into the Aldan, and were ferried over to the habitation of a Yakut (who has charge of the ferry, in company with a Coffack), 237 verfts from Yakutfk. The face of the country hitherto was undulated, confifting of wood, meadows grown with fine grafs, and an immenfe diverfity of flowers, romantic lakes, fome of them very extenfive and replete with iflands ; and here and there a folitary hut, the habitation of fome herdfman. We faw and fhot a great number of wild fowl, ducks, curlews, &c. From the Anga to the river Aldan, we found the country rather mountainous, more woody, and bearing lefs grafs. Here we arrived on Thurfday the 18th. The weather had been extremely hot and fultry all the way from Yakutfk ; but we had here a thunder ftorm, which cooled the atmofphere.

The habitations on the north fide of the Aldan confift of four huts, one belonging to Coffacks who guard the ferry, the reft to Yakuti with horfes. 331 verfts from Yakutfk the river is 500 fathom wide, flowing to the weft, and well ftocked with fifh, as are alfo the neighbouring lakes. The woods abound in wild beafts and game ; and the plains are inhabited by very opulent Tartars, who poffefs immenfe numbers of horfes and herds of cattle. The fouth of the river is bounded by perpendicular

moun-

mountains 70 fathom high, compofed at the bafe of a hard fandy ftone replete with petrified trees, very large, all lying one way, the roots north-weft, and the tops in the mountains fouth-eaft. Above there is a ftratum of loofe fand-ftones and fmall calcareous fhells, with foft greenifh earth that has a tafte like copperas and a ftrong fmell, and balls of fulphur. A ftratum follows of a much harder ftone, very compact, and impregnated with oyfter, fcollop, and other larger fhells. Another ftratum of petrified fea weeds, and wood; and then a ftratum of fmall mufcle fhells in a very hard and compact greyftone, fome of which, when broken, are found to be full of fine cryftals, &c. This mountain is fix verfts in length, and about 500 from the neareft fea. Here we found and left the firft detachment, to divide into fmaller parties.

Captain Billings refolved on profecuting his journey to Ochotfk with all poffible fpeed; to accomplifh which, he left all his baggage with the party here, and only took neceffaries for the road; and in the evening of the 19th we left the Aldan with twenty-one horfes. Our party confifted of Mr. Robeck, Dr. Merck, and myfelf; a petty officer, two guides, and an interpreter. We now left the fertile and inhabited plains, and got intangled in woody mountains and moraffes; rivers and torrents rufhing down the mountains, and all the productions of nature changing, except the larch and firs, which were now ftunted and ill grown. We found large fpots covered with wild rhubarb and rofemary; Rhe Rhaponticum and Ledum Paluftre; juniper, brufh-wood, pinks, thyme, &c. The climate equally altered, the air becoming cold and raw.

We arrived at the ford of the Belia Reka (White River) on Sunday the 21ft of June, at two P. M.; but found too much water to

justify

justify an attempt to crofs it. We therefore placed a mark in the river, and, obferving that it was falling flowly, pitched our tents, and the next morning perceived but very little alteration in the river.

Captain Billings, impatient to proceed, tried in many places to crofs ; at laft his horfe got into deep water, and he was forced to fwim over. The river is about 200 fathom wide, and the ftream was about feven knots, at the commencement of a rapid, over large ftones ; but the deep part was not above 15 fathom over. I fent him a Yakut guide and an interpreter, with fome dried bread, brandy, tea and fugar, and my fmall tent. Dr. Merck attempted to crofs ; but the rapidity of the ftream made him fo giddy, that he fell off his horfe where the water was only knee deep, and returned. A boatfwain's mate, Kopman, volunteered, whom I permitted to go with a fecond guide, fome fpare horfes, and bed clothes ; and lucky it was that I fent a guide with him, or he and his horfe would have gone down with the current. As I was very well mounted, I alfo fwam acrofs fafely, and made a good fire, at which we dried and refrefhed ourfelves, and proceeded. Our party was now reduced to five horfemen in all, and nine horfes *, two grey hounds, and a pointer. Our ftock of provifions confifted of twenty pounds of dried bread, two bottles of brandy, with a little tea and fugar, for a journey of 600 verfts through an uninhabited country. We experienced great advantage from travelling with few horfes, as we had lefs difficulty in getting forward ; and a very fmall plat of grafs at once afforded us a halting place for the night, and furnifhed food for them.

* Captain Billings left one of the guides with directions for the other parties.

On

On the 23d we croffed a very confiderable mountain called the Tfhakdall; the afcent of which was chiefly up a torrent rufhing from its fummit over large ftones. Here we obferved for the firft time the plant called by the Ruffians Piania Trava (Rhododendron Chryfanthum), held in great eftimation by all the different tribes of Siberian Tartars, as alfo by the Ruffians, for its efficacy in curing rheumatic complaints and old ulcerated wounds, from whatever caufe they fpring. It is drank in a ftrong decoction in a vapour bath, and the wounds are wafhed with it. The mountain tops are covered with this plant and with the (Pinus Cembra) creeping cedar.

On Wednefday the 24th of June, at feven A. M. we had a heavy fall of fnow, which covered the mountain tops. Not expecting fuch a fudden change from the extreme fultry weather that we had experienced a few days ago, we were quite unprovided for it, having left all our clothing, except nankeen jackets and trowfers, behind with the baggage; at the fame time the moraffes and rivulets prevented our going on foot.

In croffing a very boggy fpot our guide fung a melancholy fong, which was thus interpreted: " This is the fad fpot that was moiftened by the tears of the venerable Tfhogonnoi. The worthy old man! moft fkilful in the chace, and the conftant protector of his friend. 'Twas here that, unable to fupport the fatigues of the journey, his companion, his friend, his laft horfe, fell. He fat down by the fide of his laft horfe, and vented his anguifh in the bittereft of tears: Yes, the bittereft of tears; for he never failed in the duties of a Sochalar *. He never deferved to weep.

* The Yakuti call themfelves Socha, and the fingular is Sochalar.

The

(The third day he was relieved by a traveller and affisted home)."
The name of the place is Tshogonnoi Utabyta (the tears of Tshogonnoi). Having croffed this bog, we afcended the mountain
Unechan, on the top of which, 178 verfts from Aldan, we had a
fhower of fnow, and were quite benumbed with cold. We crept
under the Pinus Cembra, made a fire, heated fome water with
brandy, and refrefhed all hands. The clouds foon funk below
the mountain, and we had a clear fky in defcending its fouth
fide, where the fun dried our clothes. Here our greyhounds ran
down feveral hares, which the pointer brought us. A torrent
rufhes down this mountain flowing about 10 verfts. The Sam-
mach meets another from the Seven Mountains called the Kunkui,
which is the fource of the Belia Reka, winding weftward round
the mountains.

On the 25th we croffed the Seven Mountains fo called, though it
is only one with feven fugar-loaf tops; but the fummit of each muft
be croffed, on account of the immenfe ravines on each fide. We
arrived at three P. M. at the river Alachune, were ferried over,
and pitched our tents. About three verfts off are two folitary
huts, inhabited by a Coffac and Yakut, to guard the ferry, and
to keep fix hörfes for couriers. We are here 230 verfts from the
Aldan. Mountains, bogs, rivers, and fields of continual ice, fucceed each other to Yudomfky Kreft, 200 verfts. Here are three very
good huts inhabited by Coffacs on the river Yudoma; alfo Government ftorehoufes, where fupplies are kept for Ochotfk till
winter, when they are forwarded on fledges with dogs. We obtained here a little bread and fome meat; but, our horfes being in
very bad condition, we left our boatfwain's mate to wait the arrival of the firft party. We alfo left our dogs, which were fo
tired that they could not poffibly proceed any further with us.

F   Moun-

Mountains and rivers continue to Urak Plotbifha, 90 verfts. This was the place where formerly boats were built to convey the heavier articles of Captain Bering's expedition down the Urak to the fea of Ochotfk and the port. Five miferable huts inhabited by Coffacs, and Government ftorehoufes, occupy a fmall place at the foot of barren mountains, from 20 to 60 fathom high, replete with Calcedoni, or what Mr. Laxman calls the Urak onyx, in a mother of greenifh and reddifh grey ftone. The neighbourhood is extremely mountainous, containing a great variety of agates; and many large ftones in the river poffefs petrifactions and impreffions of large fhells. Some mountains appear compofed of jafper or jade.

CHAP.

CHAP. IV.

Arrival at Ochotſk.—Preſent State of that City.

WE arrived at Ochotſk on Friday the 3d of July, (computing the diſtance about 1020 verſts from Yakutſk,) and immediately ſaw Mr. Saretſheff, who informed us that he could not find timber fit for ſhip-building nearer than 70 verſts up the Ochot; and that he had only two days before ſent the ſhip-builders with 47 hired and government men to ſelect and fell proper trees as near the river-ſide as poſſible.

On the 8th July, the tranſport veſſel arrived from Izſhiga, ballaſted with black petrified wood exactly reſembling pit-coal; but fire had no effect upon it.

On Sunday the 12th our baggage arrived, with Meſſrs. Robeck and Merck, all ſafe. They croſſed the White River the evening of the day after our departure without much difficulty, the waters having much abated.

Toward the evening of the 14th, appearances indicated a fine ſucceeding day, according to the prediction of the Lamuti, who waited on the commandant, requeſting his permiſſion to allow them, the Yakuti, and as many of the inhabitants as were willing, to go the next morning on a duck-chaſe out to ſea, and return with the flowing tide. The permiſſion was made public.

F 2   Wed-

Wednefday the 15th, between three and four o'clock in the morning, the weather being calm and cloudy, about 50 fmall canoes, with Lamuti, Yakuti, and a few Ruffians, went out to fea, and returned with the tide at noon, driving before them an immenfe number of the fea-duck, called Turpan. When they were got into the bay of Kuchtui, about a mile from its difcharge into the fea, they were furrounded by more than 200 canoes, drawn up in a regular line, forming a crefcent. Thus inclofed, the tide left them in about fix inches water, and all the canoes were aground. A fignal officer (the policy mafter) appointed by the commandant gave the word for a general attack, when a fcene of the moft whimfical confufion enfued. Men, women, and children, plunged in an inftant into the water; fome armed with fhort bludgeons, and others with ftrings and nets. While one knocked on the head all that came in his or her way, others of the fame party ftrung or netted them, all hurly burly, huddling over each other. No field of battle is fubjeft to fuch a variety of incidents and tranfitions. An ill-directed blow fometimes lights on the hand of a friend, inftead of the head of the foe. Suddenly the fhrieks, fcolding, and fwearing of the women, and wrangling among all, change to peals of laughter and merriment; and the fupplication of the ducks, and the noife of myriads of gulls hovering about, form the ftrangeft medley of founds, perhaps, that were ever heard. The women caught by far the greater quantity; and the whole number deftroyed amounted to more than fix thoufand five hundred.

The Turpan is as large as a domeftic duck. The neck fhort; the bill black, fhort, and narrow, with a callous knob on the noftrils. The feathers black, with dark grey fpots. They moult all the quill feathers at once, and confequently cannot fly; being

driven,

driven, therefore, into fhallow water, they are prevented from effecting their efcape by diving, and become an eafy prey. They tafte very fifhy, but make an agreeable change of food for the poor inhabitants. When falted and fmoke dried, they are efteemed an excellent whet, with a dram, before dinner.

In the evening, a merchant's veffel, belonging to Grigory She-likoff, under the command of Yefftrat Delareff, a Greek, arrived from the Aleutan iflands, and the north-weft coaft of America, laden with furs. He left Ochotfk in the month of July 1781; arrived on the 10th Auguft at Bering's ifland, where he paffed the winter; the fecond winter he fpent at Oonalafhka; the winter of 1783 at Prince William's Sound, and the years 1784 and 1785 at Unga, an ifland off Alakfa.

Captain Billings's inftructions recommended his travelling from Ochotfk in the tranfport veffel to Izfhiga; to crofs the country of the Tfhutfki, and defcend the river Omolon to the Kovima; but it appeared not practicable for more than two or three in company to go this road, which was rendered ftill more uncertain, owing to the natives being on a doubtful footing of friendfhip with the Ruffians. This intelligence was confirmed by reports to Government brought by the tranfport veffel that arrived on the 8th inftant.

Major Shmaleff, now in this town, was the commanding offi-cer at Izfhiga, or Izhiginfk, where he was efteemed to a degree of adoration by the favage neighbours, when a Lieutenant Polo-foff was fent thither to fuperintend the difcipline of the few fol-diers in the garrifon. This man preferred feveral fecret charges againft the major, who was in confequence ordered to the government

vernment city of Irkutſk to anſwer them; but illneſs has prevented his going farther.

The Tſhutſki and Koriaki, arriving at Izſhiga after his departure, made inquiries after their old friend. Receiving no
ſatisfactory accounts, however, and not meeting with the treatment that they were accuſtomed to, they refuſed the payment of
their tribute to the new officer, who inſulted and threatened them.
They therefore left the place in the night in great diſguſt, vowing
revenge againſt the deſtroyer of their protector, as they called
Shmaleff.

The very heavy complaints on all ſides againſt Poloſoff induced
the chancery of the Port of Ochotſk, under whoſe juriſdiction Izſhiga ſtands, to order him away with the tranſport veſſel ſent there
laſt ſpring, and to leave the command of the place to a ſerjeant;
whoſe reports ſoon arrived, repreſenting the neceſſity of Major
Shmaleff's return to ſet matters to rights, and appeaſe the wrath
of the ſavages, who would not permit a Ruſſian to go among
them.

Poloſoff went immediately to Irkutſk, and, having ſpent all his
money, inſinuated himſelf into the acquaintance of Captain Billings, who, on a ſuggeſtion of the governor's, that he might be of
ſervice, engaged him in the expedition, and brought him back to
this place.

An old man now reſiding in this town, a native of the Kovima,
Lobaſheff, who has accompanied ſeveral expeditions to the Icy
Sea, repreſented an eaſy way to the Kovima by the Amicon, and
offered to guide Captain Billings; aſſuring him, that the roving

Tungooſe

Tungoofe on the road would put him to rights fhould he err. The Lamuti or Tungoofe here confirming this intelligence, the refolution was taken; and on the 22d July Captain Billings demanded 93 horfes for his conveyance, with baggage, &c. On the 25th, a courier was difpatched to the Amicon to procure a change of horfes.

On the 27th, the laft of our parties arrived from Yakutſk, with the baggage in pretty good condition.

On Thurſday the 30th of July, the commandant of this port (Lieutenant-Colonel Kozloff Ugreinin) prefented an official paper to Captain Billings, reprefenting the variance between the Ruffians and the natives near the fort of Izſhiginſk, and recommending Major Shmaleff (now here on account of the falfe reprefentations made by Lieutenant Polofoff), to be fent to Izſhiga, on the part of the expedition, to regain the friendſhip of the Tſhutſki and Koriaki.

Major Shmaleff, on application, immediately offered to go, with great affability, and in full confidence of fuccefs. He is near 60 years of age; rather above fix foot high, and very ftout; but has been fome time ill, and is ftill rather infirm. His difpofition is mildnefs and good nature itſelf; and no man in the world ever bore a better character. He offered to fail with the tranfport veffel now ready for Izſhiga; and was directed to fecure two interpreters and two Coffacks of the Izſhiga command, and, after pacifying the natives, to proceed to Virchni Kovima, and join us as foon as poffible; which, he faid, he fuppofed would be about next March. He was fupplied with money for his expences, and trinkets for prefents.

Captain-

Captain-Lieutenant Hall now received directions to superintend the building of ships here, instead of Captain-Lieutenant Saretsheff, who was ordered to accompany us. He departed with Mr. Bakoff and the party for the Kovima on the 1st August; Captain Billings, with Dr. Merck, Mr. Robeck, myself, and a few attendants, meaning to follow in two or three days.: but before I leave this place, I beg leave to lay before the reader the following account of its situation.

The city of Ochotsk is in latitude 59° 19′ 45″, longitude 145° 16′; built on a neck of land five versts long, and from 15 to 150 fathom wide, and its direction due east. It is chiefly composed of sand, shingles, and drift wood, the whole thrown up by the surf. The sea bounds it to the south, the river Ochot to the north, and its estuary terminates the neck of land. The town occupies the space of about one verst in length, contains 132 miserable wooden houses; a church and belfry; several rotten storehouses; and a double row of shops, badly stocked with handkerchiefs, silks, cloth, leather, and very bad trinkets; hams, butter, flour, rice, &c.

The air is unwholesome in the extreme, as fogs, mists, and chilling winds, constantly prevail, which so much affect the products of the earth, that nothing grows within five versts of the sea. Here stunted and withered larch-trees commence scantily; they thicken at 10 versts; at 15 versts a ridge of inconsiderable mountains are crossed, which seem to stop the effects of the sea air; for trees become more sizeable, meadows not unfrequent; and a few indigent Yakuti live hereabouts, with a very small number of horses and cows, chiefly belonging to the inhabitants of Ochotsk; who, except two priests, and the officers of the

courts

View of the Port of Ochotsk.

Published March 2d. 1801, by Cadell & Davies Strand.

courts of juſtice, conſiſt of ſailors, Coſſacks, and their families, the moſt drunken ſet I ever ſaw; but, even in their exceſſes, obedient, and never inſulting to their ſuperiors. The ſcurvy rages here with great violence, owing, perhaps, as much to want of attention and cleanlineſs in the inhabitants, as to the climate.

Fiſh is the chief food; but the finny tribe appear late, the firſt glut of ſalmon aſcending the rivers at the latter end of June. Both men and women are employed in the fiſhery, which they practiſe with a net about 20 feet long, and three or four broad; one part ſunk with ſtones, while the other is kept afloat by pieces of the bark of poplar. The net is puſhed into the river, by means of a long pole, while the fiſher ſtands on the beach. One perſon ſometimes has three nets, and catches during a tide from eight to twelve hundred. When fiſhing is over, they ſit down on the beach, ſplit the fiſh, and hang them up to dry for a winter's ſtock for themſelves and their dogs, which are uſed for draft, and each houſeholder on an average keeps twenty.

Every ſpring is a time of ſcarcity of proviſions; the dogs then become ſo ravenous, that it is not uncommon for them to deſtroy one another; and the firſt horſes that arrive are generally torn to pieces.

On our arrival, we dined with Mr. Saretcheff on cold roaſt beef, which taſted ſo fiſhy, that we thought it had been baſted with train oil. In the afternoon we drank tea at the Commandant's: this alſo taſted of fiſh; and when I mentioned it to our hoſt, he recommended the next cup without cream, which was very good. He told me, that the cattle had been fed for the laſt ten weeks entirely upon the offals of fiſh, and that the cows preferred dried ſalmon to hay.

G   The

The harbour is formed by the river Ochot, flowing from the weft and north-weft into the fpacious Bay of Kuchtui, 14 verfts long, and about four wide; fhallow, and more than three-fourths dry at low water. A river that gives name to the bay falls into its north-eaftern extremity under the Mariakan mountains. Thefe two rapid torrents, at their junction, are continually fporting with the banks, removing fome, enlarging others, and throwing up new ones: the beds of the rivers are compofed of loofe pebbles, from about the fize of a pigeon's egg to that of a fwan's. The main channel of the Ochot is only navigable for fmall empty veffels one mile upwards; for in many places the depth is only one-half to two and a half feet, or fix to eight feet at high water.

The communication with the fea has the appearance of an artificial cut 250 fathom wide, about 150 long, in a direction due fouth, and from fix to feven fathom deep: the current three to five knots ripling. The fudden check that the ftream receives from the fea is the caufe of a bank in the form of a crefcent, fouth fouth-weft, and weft, the diftance of a mile and half out: a bar continues to extend weftward, leaving a channel five feet deep at low water only, 30 fathom wide, but frequently fhifting; and this is the only navigable paffage. A very violent furf conftantly breaks over the bar, and all along the fhore. At the time of the equinoctial gales, the fpray wets the houfes of the town, and feems to threaten the deftruction of the whole place. Shoal water continues out to fea; and at the diftance of eight miles, the depth is only 10 fathom, with a bottom of loofe ftones, not compact enough to hold an anchor againft even a moderate breeze of wind.

Such is the picture of a place where we purpofe building two fhips of 260 to 800 tons burthen!

CHAP.

CHAP. V.

Departure from Ochotſk.—The Tungooſe deſcribed.—Amicon.—
Tarin Urach.—Zaſhiverſk.—Virchni Kovima.

In the evening of the 3d of Auguſt we left Ochotſk, and travelled about eight verſts, when we halted. Kopman, the boat-
ſwain's mate, who ſwam over the Belaia Reka to Captain Billings,
in croſſing a ſmall branch of the Ochot, which led to our halting
place, fell from his horſe, and was drowned; nor could we then by
any means find his body. It was, however, diſcovered a few
days afterwards by a Yakut; the head much bruiſed, and a fowl-
ing piece, which was ſlung over his ſhoulder, bent: probably in
his fall he had got entangled with the horſe's legs. The next
day we came to the ſpot where our people were cutting timber,
near Mundukan, a branch of the Ochot, in a grove of ſizeable
larch.

On the 7th, at noon, we forded the Ochot, and arrived at the
diſcharge of the river Ark, among the ſummer habitations of the
Tungooſe, who treated us with berries, and the women entertained us with a dance.

On the 9th we croſſed the Aglikit, on the borders of which, at
the foot of a mountain called the Ooyega, are ten ſummer huts inhabited by Tungooſe. Captain Billings, deſirous of getting forward with all poſſible ſpeed, obtained from them 22 rein-deer,
and halted the next day to refreſh our horſes; the neighbourhood

producing plenty of good grafs: we are here 200 verfts north north-weft of Ochotfk.

On the 11th I received difpatches for Captain Lieutenant Be-ring, at Virchni Kovima, defiring him to proceed from thence to Seredni, and collect timber to build three floops for navigating the Icy Sea; if, upon enquiry, there appeared a probability of procuring wood; if not, to ftay at Virchni.

Having with me the fhip-builder and my fervant, at three P. M. I left the party, mounted on a beautiful young rein-deer; the faddle placed on its fhoulders, without ftirrups; no bridle, but a leather thong about five fathom long tied round the head of the deer; this is kept in the rider's left hand, that he may prevent its efcape if he falls, and, when refrefhing, have a little fcope to fe-lect its food. A ftrong ftick about five feet long affifts the rider to mount; though the Tungoofe, for this purpofe, ufe their bow; ftanding on the right fide of the deer, they put the left leg upon the faddle, lean on the ftick with the right hand, and fpring up with aftonifhing apparent eafe: we, however, could not effect it by any means without affiftance; and, during about three hours travelling, I dare fay that we fell near twenty times. The top of the faddle is fquare and flat, projecting a few inches over the fides of the deer; the feat is fecured by drawing up the calves of the legs toward the thighs, and clinging faft to the projecting parts of the faddle, which at firft caufes aftonifhing pain to the thighs: by the third day, however, I became a very expert rider; the fhipbuilder could not manage it at all, and went for the moft part on foot; of courfe my travelling was not very expe-ditious.

On

Drawn by W. Alexander

View of the Peak and Rivulet of Shilcap near Ochotsk with Tungoose Tents.

Published March 2d. 1801, by Cadell & Davies, Strand.

Engraved by J. Powell

On the 16th of August we croffed a very lofty mountain, called the Oorakantfha, about half way up the afcent of which is a plain, with an extenfive lake. The paffage over this mountain is very difagreeable, up a ravine, down which a noify torrent takes its way among maffes of ftones, with tremendous overhanging rocks. We laboured twelve hours in croffing it, and found it extremely barren, not producing a blade of grafs; but in one place was a fmall bed or two of mofs, where we refrefhed our deer for about a quarter of an hour. This is one of the Vircho Yanfki chain of mountains, remarkable for being the fource of the Ochot, Indigirka, Iydoma, and Mayo rivers, and diftant from Ochotfk 415 verfts. I compute its fituation from the reckoning that I kept of courfe and time, latitude 62° 10′, longitude 144° eaft of Greenwich.

This chain has a direction nearly eaft and weft, extending about two degrees north and fouth; though fome branches appear from the latitude of 61° to nearly 67° north.

On the 17th I overtook the courier that was difpatched the 25th of July from Ochotfk, and foon difcovered that he had been making a trading trip among the Tungoofe. I therefore fent him to feek for horfes among the neighbouring Yakuti, agreeable to his orders.

On the 20th I arrived at the Amicon, which is the name of the chief fource of the Ingigirka, on the borders of which are built two Ruffian Irbas, inhabited by Coffacs, who are charged with the care of poft-horfes, or rein-deer, for travellers, this being the winter road from Yakutfk to Ochotfk; latitude 63° 5′, longitude 145°.

8

This

This neighbourhood contains, within the circumference of 30 verſts, about 20 inhabited Yakut huts. The face of the country is moderately level for about 90 verſts, interſperſed with meadows and groves of larch, poplar, aſp, birch, and alder, with underwoods of black and red currants, dog-roſe, and raſpberry. The ſituation is well calculated for the Yakuti; for, placed beyond the reach of intruding viſitors, they paſs their time in ſavage indolence, and, like the bears, their neighbours, are only rouſed from their lethargy by the abſolute calls of nature, when they prowl about in queſt of animals. The verges of the mountains that bound the plains are infeſted with bears, lynxes, wolves, foxes, elks, deer, hares, ſquirrels, and ſome ſables. The meadows ſupport their immenſe numbers of cattle and horſes, and the rivers and lakes abound in fiſh and wild fowl; ſo that a trifling effort is ſufficient to ſupply every want that they are ſenſible of.

I took leave of my Túngooſe and their rein-deer, and declare that I did ſo with regret; for I was now an adept in riding, and found them more eaſy and agreeable than horſes; but, above all, I was enchanted with the manly activity of my guides, their independence, and contentment. Satisfied with the limited productions of nature, where nature itſelf ſeems to forbid the approach of mankind, their aſtoniſhing fortitude, keeping in full force every lively ſenſation of the mind, and ſurmounting all difficulties, until they obtain the intereſting object of their purſuit, inſpired me with an ardent deſire to participate in their dangers and delights. I pronounce them " great Nature's happy commoners ;" for they are

————— " As free as Nature firſt form'd man,
" Ere the baſe laws of ſervitude began,
" When wild in woods the noble ſavage ran."

The

The romantic defolation of the fcenes that frequently furrounded me, elevated my foul to a perfect conviction that man is the lord of the creation. I confidered the dependence of the inhabitants of great cities, neceffitated to fupply the luxurious, opulent, but ftill more dependent, as the greateft and moft fubmiffive humility that refinement can impofe upon man, checking his hofpitality, and baffling all his hopes of mutual and reciprocal enjoyment, which is the bafis of fociety, and the only fource of happinefs.

The Tungoofe wander over an amazing extent of ground, from the mouth of the Amour to the Baikal Lake, the rivers Angara, or Tungoofka, Lena, Aldan, Yudoma, Mayo, Ud, the fea coaft of Ochotfk, the Amicon, Kovima, Indigirka, Alafey, the coaft of the Icy Sea, and all the mountains of thefe parts ; conftantly on the look-out for animals of the chafe. They feldom refide more than fix days in one place, but remove their tents, though it be to the fmall diftance of 20 fathom, and this only in the fifhing feafon, and during the time of collecting berries in fuch folitary places as are far diftant from the habitation of Coffacs *. Here they leave their fupplies of dried fifh and berries, in large boxes built on trees or poles, for the benefit of themfelves and their tribes in travelling during the winter. Berries they dry by mixing them with the undigefted food (lichen) out of the ftomach of the rein-deer, making their cakes, which they fpread on the bark of trees, and dry upon their huts in the fun or wind.

They feem callous to the effects of heat or cold; their tents are covered with fhamoy, or the inner bark of the birch, which they render as pliable as leather, by rolling it up, and

* They fay, that their tents contract a difagreeable fmell from remaining long in one place.

keeping

keeping it for fome time in the fteam of boiling water and fmoke.

Their winter drefs is the fkin of the deer, or wild fheep, dreffed with the hair on; a breaft-piece of the fame, which ties round the neck, and reaches down to the waift, widening towards the bottom, and neatly ornamented with embroidery and beads; pantaloons of the fame materials, which alfo furnifh them with fhort ftockings, and boots of the legs of rein-deer with the hair outward; a fur cap and gloves. Their fummer drefs only differs in being fimple leather without the hair.

They obtain fupplies of food from the Ruffian inhabitants of the Amicon, Indigirka, Uyandina, Alafey, Kovima, Zafhiverfk, Ochotfk, &c. They are religious obfervers of their word, punctual and exact in traffic; fome few are chriftened; but the greater part are Demonolatrians, have their forcerers, and facrifice chiefly to evil fpirits.

An unchriftened Tungoofe went into one of the churches at Yakutfk, placed himfelf before the painting of Saint Nicholas, bowed very refpectfully, and laid down a number of rich fkins, confifting of black and red foxes, fables, fquirrels, &c. which he took out of a bag. On being afked why he did fo, he replied, " My brother, who is chriftened, was fo ill that we expected " his death. He called upon Saint Nicholas, but would have no " forcerer. I promifed, that if Nicholas would let him live, I " would give him what I caught in my firft chafe. My brother " recovered, I obtained thefe fkins, and there they are." He then bowed again, and retired.

They

They commonly hunt with the bow and arrow, but some have rifle-barreled guns. They do not like to bury their dead, but place the body, dreſſed in its beſt apparel, in a ſtrong box, and ſuſpend it between two trees. The implements of the chaſe belonging to the deceaſed are buried under the box. Except a ſorcerer is very near, no ceremony is obſerved; but in his preſence they kill a deer, offer a part to the demons, and eat the reſt.

They allow polygamy; but the firſt wife is the chief, and is attended by the reſt. The ceremony of marriage is a ſimple purchaſe of a girl from her father; from 20 to 100 deer are·given, or the bridegroom works a ſtated time for the benefit of the bride's father. The unmarried are not remarkable for chaſtity. A man will give his daughter for a time to any friend or traveller that he takes a liking to; if he has no daughter, he will give his ſervant, but not his wives.

They are rather below the middle ſize, and extremely active; have lively ſmiling countenances, with ſmall eyes; and both ſexes are great lovers of brandy.

I aſked my Tungooſe, why they had not ſettled places of reſidence? They anſwered, that they knew no greater curſe than to live in one place, like a Ruſſian, or Yakut, where filth accumulates, and fills the habitation with ſtench and diſeaſe.

They wander about the mountains, and ſeldom viſit ſuch plains as are inhabited by the Yakuti; but frequently reſort to the ſolitary habitations of the Coſſacs appointed to the different ſtages, as they are there generally ſupplied with brandy, needles, thread, and

H   ſuch

such trifles as are requisite among them and their women, who always accompany them in their wanderings.

August 20. Immediately upon my arrival at Amicon, I sent for five horses for my party and baggage, including one for the guide, and also begged that a change might be procured for the party with Captain Billings. I was informed, that a Sinboyarsk of Yakutsk, who accompanied some of Siberakoff's contracted provisions, had obtained two days ago 63 horses; and that upwards of 200 had been lately sent from this neighbourhood to assist the party from Yakutsk under the command of Captain-Lieutenant Bering; so that but very few remained; and I concluded that the party which I left would be but badly supplied.

On Friday the 21st August, at noon, I obtained five horses, and proceeded on my journey. On the 23d, in the morning, I arrived at a place called Tarin Urach, an extensive plain, replete with lakes and woods, the habitation of several Yakuti. Here I found the Sinboyarsk from Yakutsk, with Siberakoff's provisions. His name was Ivan Yefimoff; and he, with the inhabitants, persuaded me to float down the Indigerka on a raft, to the habitation of the Yakut prince, Nicolai Samsonoff, where I could be supplied with horses, and proceed on a good and strait road to Virchni Kovima. They assured me, that it was the shortest road, not obstructed either by rivers or mountains; and that such travellers as required but few horses always took this route, which did not produce grass enough for caravans, or great parties.

I was offered four men to take me to the Yakut prince's; and told, that, if I chose it, they would get me two small rafts immediately.

I

mediately.

mediately. I agreed to this, and the next morning, at nine o'clock, embarked, made 70 verfts by dark, and pitched my tent in a wood near fome tremendous rocky mountains. The night was windy and rainy; and the howling of wolves at no great diftance prevented our getting much fleep. In the afternoon of the 26th, we arrived in the neighbourhood of the prince's habitation, to which we walked, and found him extremely drunk; fo that it was with difficulty I obtained two horfes to fend for my baggage.

The next morning, at a very early hour, I awakened the prince, who apologifed for having been drunk, declared that he had no horfes at home, nor any man except an old fellow; and that 60 horfes and all his men had been fent about ten days ago to Captain Bering's party at the Momo. He told me, that the feafon was too far advanced for me to travel the road pointed out; but that there was a probability of fuccefs on the way that Mr. Bering had taken from the river Momo.

Friday the 28th, at nine o'clock in the morning, I obtained horfes, and immediately proceeded on my journey to the Momo, which we croffed on Sunday the 30th; the country being generally level, with abundance of brufh-wood. This morning we faw upwards of 20 hares, and arrived towards evening at the habitation of an unchriftened Yakut chief, named Choratin, a very hofpitable man, who faid that Captain Bering had paffed the Momo on the 16th, and loft feveral horfes in croffing the river; but that now this road was not paffable, and no other way was left for me to go than through the town of Zafhiverfk, whither he would conduct me himfelf.

H 2   I was

I was extremely forry to get fo much to the north-weft of the place of my deftination; but, as there was no mode of avoiding it, I was compelled to fubmit.

We profecuted our journey the next morning, and arrived at Zafhiverfk on the 3d of September at noon. I made immediate application to the mayor (Mr. Samfonoff) for his affiftance, not only regarding my travelling the beft way, but with refpect to provifions, mine of every kind being completely exhaufted in the morning, and I hoped here to procure a fmall fupply. Mrs. Samfonoff gave me fome tea, fugar, and bread, out of a very fmall ftock of their own. They were very happy to fee a European, the firft (except a general in exile) that they had beheld for four years; and their behaviour was extremely polite and kind.

This town contains one church, five ifba's, or Ruffian houfes, and 21 huts, on a boggy point of land running into the Indigirka. The oppofite fhores are barren perpendicular mountains, producing in ravines here and there a ftunted larch-tree, as defcribed in the annexed ENGRAVING. Its fituation I compute in latitude 66° 30' north, longitude 142° 10' eaft. The inhabitants confift of the mayor and his wife; the captain of the diftrict and his wife, now refiding (for the fake of fifhing) 40 verfts down the Indigirka; two priefts, brothers, and their attendants; two writers; and all the reft are Coffacs.

The mountains embay the town eaft, fouth, and weft; fo that the fun is only vifible three hours and 30 minutes at this feafon; from the 12th November till the 6th January O. S. it is hid, and the place is enveloped in night.

On

Page 82.

View of the Town of Lashwersh.

Published March 2d 1802, by Cadell & Davies, Strand.

On the 4th, at five P. M., we left Zafhiverfk upon the fame horfes that brought us, but with two guides. The next day we croffed the Indigirka at the ferry called Samondran, 40 verfts from the town. A little to the north of this place we obferved the branch of the Virchoyanfki chain terminate by low and detached mountains; the Arctic flats fucceeded, which are very boggy, except here and there an elevated fpot producing a clufter of ftunted larch-trees, oziers, and alders; the other parts are occupied by an immenfe chain of lakes, all joining by narrow runs.

On the 13th we arrived at three inhabited Ruffian ifba's on the river Uyandina, near its difcharge into the Indigirka, about the latitude 67° 45', and longitude 148° 35'. Thefe people carry on a trifling trade with the wandering Tungoofe, and the Yakuti, that go in queft of mammont's tufks, giving in return dried fifh, and flour, with fome articles of drefs and ornament.

Near this place we recroffed the Indigirka, and travelled nearly eaft to the Alafey mountain, which I eftimate in latitude 67° 8', longitude 153° 10'; from hence our road led nearly fouth; the country became more uneven, and better wooded, to Virchni Kovima, where I arrived on the 28th September, after fuffering innumerable hardfhips in this roundabout road, and being the laft 16 days without either bread or falt, living merely on dried fifh of bad quality.

Captain Billings had arrived on the 8th, and all his party a few days after him. Captain Bering was here only four days before him; but part of his convoy were not yet arrived, though they left Yakutfk on the 16th of June. The glafs was now 18° below the freezing point of Reaumur, and all the rivers were frozen over fufficiently to fupport horfes.

CHAP.

CHAP. V.

Meeting of the Command, under circumstances of difficulty.—Reflections.—Visit the Yukagiri.—Occupations on our return.—Virchni Kovimskoi and its Inhabitants described.—Cossacs.

HOWEVER happy I considered myself on rejoining my companions, the prospect that it opened to my view was truly melancholy. No provisions had arrived, although the contractor's time for the delivery was stipulated at farthest for the 1st of August. The stock that we had was insufficient for the road; and the inhabitants, consisting of eight males, were in a miserable situation themselves; for, not expecting such a number of visitors, they had not made any preparation for them, and had only secured a scanty winter's supply of fish for their own use. To add to the general calamity, the rivers and lakes were now destitute of fish. The habitations were five half decayed isba's, and one extensive hut, besides a chasovnoi (house of prayer), which necessity compelled us to convert into barracks. Two earthen huts were immediately constructed, one for our Izshiga Cossacs, who wished to live together; the other I took, with Messrs. Main and Varonin. Two sheds also were erected; one for our instruments, &c. the other to serve as a work-shop for our shipbuilders; these were covered with sail-cloth. We also built a smithy.

Captain Billings sent all the horses that he could collect to fetch in such provision as might be found scattered about the roads and woods, by the falling of Siberakoff's horses, and at no great distance.

On

On the 22d of October, the laſt of the party, under Captain Bering's charge, with baggage, &c. arrived, which increaſed our number of working hands to 78, excluſive of Yakuti. Every article was more or leſs damaged, and many things were loſt and left behind at different places, where the pack-horſes had died of fatigue and want of food. Soon, however, we were relieved by frequent arrivals of flour and butter.

On the 26th the ſmithy was finiſhed, and a travelling forge erected, which we had brought with us. Timbers were preparing for building a veſſel of 50 feet keel, and every thing going on with the greateſt alacrity, although numberleſs difficulties were to be ſurmounted, and all hands reduced to a ſorry pittance of bread and ſalt. Notwithſtanding all this, a ſpirited and determined reſolution exiſted every where. The Yakuti within 150 verſts ſupplied horſes to drag the felled timber three verſts up the river Yaſaſhnoi to the ſheds. Wiers were made and placed in the river to enſnare the finny fry; but none made their appearance until the 29th, when 45 large nalime were caught, and the next day 60, which afforded great refreſhment.

With the month of November the weather came in almoſt inſufferably cold; the thermometer indicated from 32° to 37° and 41° below o of Reaumur; mercury proved of no uſe in meaſuring the degrees of cold beyond 32¼°; but our ſpirit thermometer never froze. I ſhall take the liberty to inſert here our thermometrical remarks for eight days, with one of Mr. Morgan's, filled with ſpirits.

Nov. 22.

			Wind.	
Nov. 22.	4 A. M.	38½	S. W.	Light airs.
	6	39¾		9 ounces ☿ frozen in 2 hours, the
	8	39¼		earth, ice of the river, timber of
	12 M.	38½		the houses, &c. cracking, with re-
	4 P. M.	39	S. S. E.	ports equal to that of a musket.
	6	39¼		
	8	39¼		10 ounces of Mercury in a stopped
Strong N. lights 12		40		phial froze in 2½ hours.
23.	4 A. M.	37½		
	6	36		
	8	32		About 10 o'clock the ☿ in a stopped
	12 M.	32	S. E.	phial thawed.
	4 P. M.	32¾		Little wind.
	6	32¾		Mercury frozen.
	8	30½		N. B. About half an hour only; during
Strong N. lights 12		33.		which time Mercury was not com-
24.	4 A. M.	34	N.	pletely thawed, and was soon quite
Light 6		35		frozen again.
Airs	8 A. M.	36		
	12 M.	35½		
	6 P. M.	35		
Strong N. lights 12		36		
25.	4 A. M.	34½	S. W.	Little wind.
	12 M.	34¼	N. W.	Ditto.
	4 P. M.	35		
	6	36		
	8	37		
	12	38		
26.	4 A. M.	39½	S. E.	Thick fog; the earth and river crack-
	8	40½		ing violently.
	12 M.	40½		
	8 P. M.	40½		
	12	41¼		
27.	4 A. M.	40½	N. E.	Very light airs.
	6	40¼		Quite calm.
	8	40		At 9 A. M. a bottle sealed with
	12 M.	38		Astracan brandy (called here French
	4 P. M.	39		brandy) exposed to the frost, thick-
	8	40		ened very much, but was not frozen.
	12	40½		

28.

Wind.

Nov. 28.	4 A. M.	37¾ E. N. E.	
	8	33 S. E.	
	12 M.	32¾	Mercury thawed.
	4 P. M.	31	
	6	30½	
	8	31½	
	9	32½	At 10° 30′ obferved ☿ frozen.
	12 M.	36¼	At 33°.
29.	6 A. M.	38¼ to 39° all day.	
30.	6 A. M.	35¼ S. S. W.	Little wind.
	8	33	
	12 M.	31½	Mercury thawed.
	4 P. M.	31½	
	8	32	
	12 N.	32	

At 37° it was almoſt impoſſible to fell timber, which was as hard as the hatchet, except it was perfectly dry; and in the greateſt ſeverity the hatchets, on ſtriking the wood, broke like glaſs. Indeed it was impoſſible to work in the open air, which compelled us to make many holidays much againſt our inclination.

The effects of the cold are wonderful. Upon coming out of a warm room, it is abſolutely neceſſary to breath through a handkerchief; and you find yourſelf immediately ſurrounded by an atmoſphere, ariſing from breath, and the heat of the body, which incloſes you in a miſt, and conſiſts of ſmall nodules of hoar ice. Breathing cauſes a noiſe like the tearing of coarſe paper, or the breaking of thin twigs, and the expired breath is immediately condenſed in the fine ſubſtance mentioned above. The northern lights are conſtant and very brilliant; they ſeem cloſe to you, and you may ſometimes hear them ſhoot along; they aſſume an amazing diverſity of ſhapes; and the Tungooſe ſay, that they are ſpirits at variance fighting in the air.

I

Our

Our fishing continued, but gradually decreased after the first four days; and with the month of November the fish nearly left us, reducing us again to bread and water. At times, indeed, a few were caught till the middle of December.

We had now and then supplies of flour arriving, and by the end of the year the quantity received amounted to 2042 poods; but, not having ovens enough to bake bread for all hands, the generality of them used to boil the flour, and eat it with fish-oil.

Toward the end of the year the scurvy made its appearance, though not in a dangerous degree, and affecting but a very small number. The cold increased to 43°, which froze our Astracan brandy. By Christmas, we had the keel laid of a vessel of 50 feet, and resolved upon building another of 36 feet, with boats. The leather bags which contained our flour were appropriated to the purpose of making a baidar; there now existing no probability of Siberakoff's delivering hides for that purpose.

Our working hands were increased by 16 Cossacs from Neizshni Kovima, sent by the commander of that place, making the number 94, exclusive of officers. The poor horses employed in dragging timber from the woods exhibited such a picture of misery as perhaps never before existed; they were fed with brush-wood and the tops of willows, having neither grass nor hay. They seldom worked longer than a fortnight, then tired and died.

Our only happiness was derived from general harmony among ourselves, and a resolution to overcome every difficulty, to secure the means of leaving this worst of all places in the world, as soon as the ice of the rivers should break up, and afford us a passage.

Animated

Animated by this spirit, notwithstanding the severity of the weather, every thing went on with amazing success. Our joy was
increased, by obtaining for the Christmas holidays a supply of
meat from some Yakut chiefs who visited us ; and it was doubled
toward the close of the year by a prospect of better times.

A man who rolls in affluence, and knows neither cares nor
sorrows, can hardly feel for those of others, and is of all people in
the world the least qualified for pious deeds.   Let him but visit
these regions of want and misery ; his riches will prove an eye-
sore, and he will be taught the pleasure and advantage of prayer.
Let the advocates for the rights of man come here to enjoy them ;
for this is the land of liberty and equality ! Nor will the Directory
of the Great Nation, with all their great generals, ever possess it
in perfection until they have reduced their country to the independent state of this part of the globe ; where a man sees and
feels that he is a man merely, and that he can no longer exist
than while he can himself procure the means of support.

Our distress, and hopes of relief from the mercy of heaven, led
us one and all to devotion on the first day of January 1787 : and
never was a fast-day in England more devoutly passed in prayer
for plenty ; for there never existed there, nor ever will, I hope,
such a scarcity.

The Yakut chief who had supplied us with horses was this day
rewarded for his attention and losses, by the present of a silver
medal, which was fastened about his neck with proper ceremony.

We had the sun at this time three hours above the horizon, yet
the cold by no means decreased ; now and then, however, we

made

made a trifling excurfion, and were charmed with the appearance of partridges and hares, which induced us to fend out our jagers for the benefit of the community; but they were not very fuccefsful.

On the 14th of January Captain Billings propofed a vifit to the Yukagiri (who refide about 50 verfts from hence), to fee their manners and cuftoms, and procure a vocabulary of their language. He was accompanied by Dr. Merck, Mr. Robeck, our drawing-mafter, and myfelf; and we were conveyed on narti * drawn by dogs.

This method of travelling did not anfwer my expectation. We had 13 half-ftarved dogs to each fledge, which contained very little baggage; and I kept pace with them, walking the greateft part of the way on fnow-fhoes. We were nine hours on the road; but about midway we made a halt to eat fome raw frozen falmon, which I thought excellent, although it was the firft time that I had ever taken my fifh dreffed by a 30° froft; nor had I any other fauce than falt and hunger.

We arrived pretty late in the evening, and put up at the hut of the chief; a man fo remarkably ftupid, that he could not tell us how many children he had till he called their names over, bending a finger to each; and, after all, they were only five daughters and two fons. The whole number of inhabitants was 27 males and 23 females, including children.

After having taken a refrefhment of tea, with bread and butter, eight of the young women of the village came to entertain us

* Thefe are a kind of long fledges, very narrow and low.

with

with fongs and dances to a ftrange inharmonious monotony of found; and their action was an uninterefting difplay of their manner of hunting, fkinning, and dreffing the fkins of animals.

The next morning we began to make our obfervations; but found that all their old cuftoms were abolifhed, and that the race was almoft extinct. They call themfelves Andon Domni, and are ignorant who gave them the name of Yukagir. They are in tribes, and, befides this place of refidence, have villages near the eftuaries of the rivers Indigerka, Yana, and Alafey. Their cuftoms were like thofe of the Tungoofe, with whom they live in great friendfhip, and fome of the tribes intermarry. The whole nation comprifes only about 300 males, as wars with the Tfhut-fki and Koriaks have fwept off great numbers, the fmall-pox ftill more; and the venereal difeafe now feems engrafted among them, as if finally to eradicate the race. They refide at thefe habitations from the middle of December till the middle of February, while the weather is too fevere for the chafe; alfo in June and July, being the fifhing feafon. They frequent the fources of the Kovima and Yafafhnoi in queft of deer and wild beafts, which they float in rafts to their dwellings, or bring in narti with dogs. They fpeak Ruffian very well, which enabled me to take a good vocabulary of their language. Their drefs is now the fame as the Ruffians of thefe parts: it was formerly like that of the Tungoofe, whofe tailors they ftill remain, embroidering the ornamental parts of their cloathing, for which they receive in return articles of drefs, fkins, or furs. The Yukagiri call the Tungoofe Erpeghi.

On the 18th January we returned to our dwellings, and found every thing going on with alacrity. We fet our coopers to ftave-making, and began building boats and one baidar.

13   At

At the commencement of the month of February, the weather during the day began to be more moderate. On the 4th, we fent a foldier to Seredni, or the Middle Kovima, to bake bread for the enfuing fummer; and a cooper to Neizfhni, to make cafks: we alfo fent Lobafhkoff, a Coffac Sotnik well acquainted with thefe parts, to purchafe the meat of rein-deer of the inhabitants or wandering tribes about the Omolon; he was furnifhed with falt to preferve it, and with money, tobacco, and trinkets, for the purpofe of barter.

Not having any agent at Irkutfk for the purpofe of difpatching the articles that ftill remained to be forwarded to Ochotfk for the expedition, it was thought neceffary to fend Captain-Lieutenant Bering to fuperintend this bufinefs. He left this miferable place on the 12th February, and took commiffions from every officer for private fupplies of neceffaries, both of food and raiment.

The fcurvy gained ground upon our people, affecting their joints, and contracting them, particularly the legs. A decoction of the Pinus Cembra was ufed, and alfo fweetwart and quafs, and with fuccefs.

In the month of March we had our veffels in a great ftate of forwardnefs, and were warping planks for fheathing. The days were pleafant, but the night-frofts continued from 20° to 32°. On the 12th day of the month the fnow-larks made their appearance, to my great joy, for they afforded me many a good dinner.

On the 1ft of April, Captain Billings pitched his aftronomical tent. On the 8th, a Yakut arrived, with 14 fmall cafks of butter, which had wintered on the road; but brought no news of

any

any more of Siberakoff's contracted articles. We were now making fails; and a rope-walk was at work by the 20th.

On the 29th, fwans were obferved flying to the north; on the 23d, geefe; on the 26th, ducks; and toward the end of the month we obtained abundant fupplies; among which, we now and then obferved a fmall-fized goofe quite white. The fcurvy entirely left us as foon as we returned to the ufe of folid food.

The 1ft of May, at four A. M. we had 22° of froft; and at eight A. M. 23° of heat in the fun. Our people were now employed in caulking and preparing rigging for both veffels, and making oars and fweeps for the boats. Some hands were fent to fhoot birds, hares, &c. which were in great plenty, and not very fhy.

All appearances were now as favourable as they had lately been difcouraging; and perfect health, good fpirits, and fatisfaction, appeared in every countenance; when an accident, for the moment, threatened the worft of confequences. On the 14th, a little paft midnight, we were alarmed by a fire breaking out at the dwelling of our mechanic, only a few yards from the fpot where our veffels lay on the ftocks, quite ready for launching; and thefe were with the greateft difficulty faved. All the brandy that had hitherto arrived for the ufe of the expedition, which confifted of 51 ankers, and was depofited in a ftore-room adjoining to the houfe, was confumed. This accident was caufed by the careleffnefs of the inhabitant, who had made a fire-place in the entry clofe to the wooden wall, where he dreffed his fupper, and left the fire burning.

On

On the 15th, the ice of the river Yafafhnoi began to move, and the following day it floated with the ftream.

On the 17th, we launched the larger veffel, and called her *The Pallas*, as a mark of the refpect we bore to the very learned Doctor of that name, who was the chief caufe of the expedition taking place; though the original fuggefter of it was the Reverend William Coxe, A. M. F. R. S. author of " An Account of the Ruffian Difcoveries between Afia and America," and of many other valuable works, too well known to require mention in this place.

The water of the river had rifen 12 feet perpendicular, and remained fo all the 18th. On the 19th, we launched the fecond veffel, which we named the Yafafhnoi, and the command of her was given to Captain-Lieutenant Saretfheff. The perpendicular rife of the water on the 21ft was 22 feet, even with the borders of the river; and the next day it overflowed the Oftrog, and compelled us to retreat to the tops of our houfes, where we pitched our tents. The baidar and two boats were now finifhed, and, carrying all the materials in them from the ftore-houfes, we loaded our veffels. The Pallas was cutter-rigged, and the Yafafhnoi had three lugs and a fore-fail.

The perpendicular rife of the water on the 24th was 27 feet. The face of the country refembled an immenfe lake, and fome of the tree tops appeared juft above the water. In the afternoon all hands went on board. Mr. Main received the charge of fuch as were not required with us, and directions to return to Ochotfk with all convenient hafte, to affift in conftructing the veffels there.

Virchni

Virchni Kovimſkoi Oſtrog is ſituated on a boggy ſpot, over-grown with willows and alder buſhes, bordering on the river Yaſaſhnoi, three verſts from its diſcharge into the Kovima. Its latitude is 65° 28′ 25″; and longitude, by ſeveral ſets of lunar obſervations, 153° 24′ 30″ eaſt; variation of two compaſſes 7° 33′ eaſt. The number of buildings that it contains I have already mentioned. The inhabitants are Coſſacs, their wives, and attendants.

A Coſſac at Irkutſk is employed, by the governor and chief officers, in the moſt contemptible drudgery, ſuch as cleaning the ſtable, ſcowering the kitchen, making fires, &c. At Yakutſk he is of more conſequence, and finds employment as tranſlator and emiſſary; but is faithleſs, ſly, and crafty. He lives in this part of the world like an independent chief, keeping Yakut labourers to aſſiſt his wife in all domeſtic drudgery, fiſhing, cutting wood, &c. Her particular province is to wait on her huſband, whom ſhe aſſiſts in putting on and pulling off his clothes, which ſhe keeps in good repair; ſhe alſo dreſſes his food and ſerves it up; and when he has made his meal, ſhe ſits down and eats with the reſt of the labourers.

Girls are frequently married to the Coſſacs at the early age of twelve; and, as it is a ſlave that they want, it ſeems a matter of indifference to them whether ſhe be Ruſſian, Yakut, Tungooſe, or Yukager, provided ſhe profeſſes the Greek faith. Both ſexes ſeem incapable of forming any tender attachment; the women are very inconſtant to their huſbands; and the worſt of diſorders is deeply-rooted among them and all their neighbours, having been introduced by Pavlutſki and his followers, who were ſent hi-

K   ther

ther to fubdue the Tfhutfki, and communicated this diforder to all the other tribes.

The lordly Coffac is only to be roufed from his indolence by an order from his fuperior; and then he curfes his fate, which has placed him under the control of others. Thefe laft of mankind, unworthy of the name, thefe hardly animated lumps of clay, exert the moft favage barbarity over their wives, children, animals, and the poor neighbouring tribes whofe miferable lot it is to pay tribute to them, or to be under the leaft obligations, either by drinking a glafs of brandy, taking a leaf or two of tobacco, or in any other way. They receive annual fupplies of articles that are neceffary, ornamental, or luxurious, from the traders at Yakutfk, to fupply the different tribes with; rendering, in return, furs and mammont's tufks. Their chief endeavour with thefe wanderers is, to get them indebted for any article that they may ftand in need of, or to procure the receipt of a trifling prefent (which in honour they muft return with one more valuable); but if they once get in debt, then they are perfecuted to the utmoft, and are frequently neceffitated to leave a man to work, or a woman, perhaps a daughter, as fecurity for the payment.

I have here fketched a faithful picture from the very men who are fent hither to explain to the natives the benefits arifing from the Chriftian faith, and to fet an example of loyalty and obedience.

CHAP.

CHAP. VI.

Departure from Virchni Kovima.—Seredni Kovima.—Inhabitants of the River Omolon.—Neizſhni Kovima.—Shalauroff's Wintering-place.—Laptieff's Mayak.—The Pallas conſecrated, and Captain Billings advanced in rank.—Paſſage much annoyed by Ice.—Spiral Bay.—Wolves' Bay.—Barranoi Kamen.—Captain Billings reſolves on declining any farther attempt to proceed, and the Command returns to Neizſhni Kovima.

MAY the 25th, at ſeven o'clock in the morning, we left Virchni Kovima Oſtrog, and falling down the Yaſaſhnoi, entered the river Kovima about eight.

It is impoſſible to give any deſcription of this part of the river, becauſe the ſhores and iſlands were overflowed. Its direction, however, is nearly north-eaſt, and the navigation was rendered extremely difficult, owing to the current in many places ſetting with great rapidity into the woods.

We arrived at Seredni Kovima on the 28th, at nine o'clock in the evening. This oſtrog contains 15 iſbas and a church; the inhabitants, though of the ſame claſs as thoſe at Virchni, are better circumſtanced, and much more induſtrious, cleaner, and healthier in their appearance; which I attribute to the ſpirit of emulation that they poſſeſs from the activity of the prieſt, who, like a good ſhepherd of his flock, attends them to their different fiſhing-places at the various ſeaſons of the year, and preſides over the diviſion;

K 2

upon

upon which occafion, however, though he were not prefent, there would not exift any difputes. Fifh are fo extremely numerous, that, had not the feverity of the weather in winter prevented any communication, we might have received ample fupplies at Virchni from hence.

This place is fituated in latitude 67° 10′ 14″, longitude, by time-keeper, 157° 10′; variation of two compaffes gave the mean, 9° 19′ eaft.

Here we finifhed an anchor, which was begun at Virchni; but the fwell of the rivers prevented our proceeding with it: we alfo took in a ftock of fifh and bread.

The weather was very variable upon our arrival, with a fouth-weft wind, and extremely hot. But it foon fhifted to the north, and on the 2d, 3d, 4th, 5th, and 6th of June, we had froft and fnow, with 4, 5, and 6° below o of Reaumur at nights; during the day-time the thermometer indicated o, and one degree below it.

The river Kovima was not yet within its limits; but on the 11th we profecuted our voyage. The eaftern bounds of the river are broken perpendicular mountains, producing in ravines a few very ftunted larch-trees. The weftern fhores are low, and in fome places ftill overflowed; but here and there an elevated fpot produces a clufter of very thin and low larch-trees. Our veffels were frequently carried aground on the overflowed iflands; but, by fending out a fmall hawfer into the main channel, and taking into the long-boat a few bags of flour, we got off. Such accidents as thefe, with contrary winds, prevented our arrival at the Omolon

4   fummer-

fummer-huts (fix in number, oppofite the river of the fame name) before the 16th in the morning. Thefe huts are 350 verfts from Seredni, and were erected for the purpofe of fifhing, during the month of June only, by the inhabitants of the river Omolon, who are exiles, and the only people of thefe parts that have European countenances. They are in number nine males, the youngeft 50 years of age, and about 12 females. They were emancipated by an act of grace about ten years ago, and pay a head-money equal to the Siberian peafants. Their employment is fifhing, feeking the tufks of the mammont, and hunting animals about the neighbourhood. They trade with the Koriaki and wandering Tungoofe ; they are, however, very poor, and pretend to be more fo than they really are, to evade making prefents to the collectors, &c.

At thefe folitary huts we found Major Shmaileff, with two interpreters for the Tfhutfki and Koriak dialects. He has completely fettled all difputes, and reconciled the above two nations with the Ruffians. They received him with great kindnefs, and affured him that they would affift the Expedition to the utmoft of their abilities, and meet him on the fea-coaft near the Tfhaoon next fummer. The Major brought with him a very great fupply of dried deer's meat, which we took on board. He embarked in the Yafafhnoi ; and we took the interpreters, Dauerkin and Kobeleff, into the Pallas.

The next day, being the 17th June, we profecuted our voyage, and arrived on the 19th at the Oftrog Neizfhni Kovima, on an extenfive ifland. Here are about 70 houfes, and a church ; alfo an oftrog, inclofing government ftore-houfes, &c. in a fquare of compact palifadoes eight feet high, with four entrances each, fupportsupporting a tower. These oftrogs are for keeping prifoners in, and alfo to ferve as places of defence, being pierced with fmall holes to point a mufket through, and thick enough to repel a fhot from the rifle-guns of the wandering tribes; latitude 68° 17′ 14″, longitude 163° 17′ 30″; variation 14° 14′ eaft.

We took in a little falted deer's meat, left the Yafafhnoi to undergo fome alteration in her rigging, and gave the prieft orders to come in her as far as to the difcharge of the Kovima, to confecrate the veffels, and to adminifter the oath to Captain Billings; who, according to the mandate of her Imperial Majefty, was to declare himfelf a Captain of the fecond rank upon his arrival in the Icy Sea.

On Saturday the 19th June, at half-paft fix in the afternoon, we weighed anchor, with a moderate breeze from the fouth-eaft. The waters were much abated; the depth of the river was 12 fathom; its width three miles; and its direction about north-eaft. Thirty-five miles below Neizfhni we obferved the laft tree; brufhwood continued a little farther.

On the 20th, at nine A. M. we arrived off the place of Shalauroff's wintering in 1762; confifting of a large ftore-houfe and double dwelling-houfe of wood in decay, under inconfiderable mountains, compofed of flate and quartz, covered with mofs; great quantities of drift wood lying on the fhore. The productions of the earth are willow and birch bufhes about eight inches high, and the diftance about 80 verfts from Neizfhni. Captain Billings, Dr. Merck, and I, went on fhore, and collected a few plants; as wolffbane, a wild vetch (the root of which is the fupport of the marmot), tanzy, and a fpecies of rock fern (the leaves not exceeding three inches in length, with an aromatic tafte, and

pleafant

pleafant fmell). Here we lay at anchor three hours. At three quarters paſt twelve we proceeded, the depth of the river decreafing gradually to one fathom. At five P. M. we agáin came to anchor, and fent out a boat to find the proper channel; the foundings varying much, fix, feven, eight, and ten feet, deepening to feven fathom. At eight we again weighed, and found the width of this branch of the river 12 miles.

Notwithſtanding our navigation was impeded by fhallows at the diſcharge of the river, where we frequently got aground, we cleared all of them about midnight, and caſt anchor oppoſite Laptieff's Mayak *, five miles from fhore, in four fathom water: fand-banks prevented our nearer approach.

On Monday the 21ſt June, at two A. M., Captain Billings took his aſtronomical tent on fhore; and our naturaliſt, Mr. Robeck, and I, accompanied him. We could not get within two miles of the Mayak on account of the fhoals, which compelled us to land in a bight two miles and a half more to the eaſt. On rowing towards fhore in a fmall boat, we got the Pallas's hull down, and fhortly after fhe difappeared; increaſing our diftance, fhe was again feen, hull, rigging, &c. feeming of an immenfe fize, and confiderably above the horizon. The weather was rather hazy, and the fun obfcured.

This morning we walked acrofs the head-land to the buildings in the next bight, which confiſt of three ifba's adjoining each other under a hill; upon which is a pyramidical building 25 feet high, fupporting a crofs, bearing the infcription " SHALAUROFF,

* Beacon, or light-houfe.

1762."

1762." The huts were built by Laptieff and his company in 1739. Where he wintered, at a small distance, is a cross bearing an illegible superscription; and a stage, about ten feet high, covered with earth, upon which they made signal fires. The shores are covered with drift wood. This is the resort of different tribes for the pestsi, or stone fox; and numbers of falls, or traps, are placed in different parts. We observed the traces of wolves, and in the afternoon two approached very nigh to our tents. Two dogs gave chase to them, but were not equal in speed.—Hazy and misty weather.

On Tuesday the 22d, at eight P. M. the Yasashnoi arrived, and cast anchor about 100 fathom south-west of the Pallas; hazy and misty weather continuing. At midnight flying clouds; and at intervals the sun visible.

On Thursday the 24th, at four A. M., we struck the astronomical tent, without having been able to take a single observation, owing to thick weather. At nine o'clock we went on board; Captain-Lieutenant Saretsheff and Major Shmaileff accompanied the priest to the Pallas, which vessel he consecrated, and after service administered the oath to Captain Billings for his advanced rank. At eleven he returned in his boat to Neizshni, and Captain Billings sent dispatches to the Governor-General of Irkutsk, with others to be forwarded to St. Petersburg.

At noon we weighed anchor with a gentle breeze at south south-west, shaping our course north north-east, the depth varying from one and one-fourth to three fathoms, and keeping a boat ahead employed in sounding. At six P. M. we saw the first ice floating near the vessel; immediately after, the wind shifted to

north

north by eaft, bringing a very thick fog. We came to anchor in four fathom, about four miles from fhore. The Yafafhnoi anchored aftern.

On the 25th, at ten A. M., a moderate breeze fpringing up from the north-weft, we again weighed, and ftood to the north north-eaft, the atmofphere very foggy. At eleven faw great quantities of ice to the north; at five P. M. were quite furrounded with ice; foundings feven fathom, fand and clay. At fix the ice compelled us to ftand-in for fhore, having run about eleven miles north-eaft. At eight, being about a quarter of a mile from fhore, off a fmall bight, caft anchor in two fathom water. We faw four black bears on the beach, manned our jolly-boat, and fent our chafers after them; but in vain. At ten Captain Billings took his aftronomical tent and apparatus on fhore.

All the 26th was hazy and wet; great quantities of ice floating and collecting to the north-eaft. Mifty weather continued all the 27th; and, on account of ice gathering about the veffel, at nine P. M. we hauled her clofe into a fmall bight in feven feet water; but were obliged at eleven P. M. to weigh, and ftand away to the weft, the only paffage open. We had a gentle breeze from the north-weft, which, frefhening on the 28th, brought ice about us. We hauled about five miles weft, and, getting into a fnug bight, dropped anchor. On account of fpiral rocks on the top of the mountains that bound the bay, we called this place Spiral Bay. At noon we had a fight of the fun; and, having fent for our tent and inftruments from our laft anchoring-place, obferved the latitude to be 69° 27′ 26″, longitude, by time-keeper, 167° 50′ 30″.

L

We

We had calms and variable light winds, with thick weather, till the 1ft of July at noon, when we again weighed, the weather being hazy, with a frefh eafterly breeze, keeping as near the eaft as poffible. At eight P. M. Captain Billings refolved to fhape his courfe north, to fee how the ice was in that direction. We obferved that the current carried us two points weft, and our foundings gradually increafed from four to 15 fathom. At midnight our rigging was covered with ice, the thermometer, about feven feet above the fea, indicated one-half above the freezing point. Thick fog ftill prevailed.

At two A. M. on the 2d, we got among very thick detached pieces of ice; which increafing upon us, our depth decreafed to nine and feven fathom. Wore fhip, and bore away to the fouth, having loft fight of the Yafafhnoi in the fog at ten laft night. The ice was not fo compact as to prevent our going farther; and from our fhoaling water, I was inclined to think that we fhould foon fall in with either the continent or fome ifland; I therefore wifhed that Captain Billings would have continued his northern courfe. The wind blew frefh; but the quantities of ice kept the fea down, and the water was quite fmooth. He was fearful of being entirely hemmed in, and was under ferious apprehenfions for the fafety of the Yafafhnoi, which was a fmall flight-built lugger. At eight A. M. we got clear, and obferved that the fog hovered over the ice only. At noon we came into a pretty deep bay (which we named Wolves' Bay, from our feeing feveral of thofe animals on the mountains); and, dropping anchor, fent three failors on fhore, with directions to proceed to the next weftern promontory to look out for the Yafafhnoi, and make a fignal-fire.

On

On the 3d of July, at four P. M., we weighed, with a moderate south-east breeze, and stood off and on three hours for the sailors on shore; when, the wind veering to north-east, we again came to anchor near the same place. At midnight we had flying clouds, the sun was visible, and a beautiful rainbow was seen in the south.

On the 4th, at five A. M., the sailors returned on board, and said that they had walked to Cape Kovima, where they made a signal-fire, having seen the Yasashnoi at anchor about ten versts off. She sent her boat on shore to inquire after the Pallas, and at six P. M. came along-side.

On the 5th, at four A. M. again weighed anchor with a wester-ly breeze; and, on account of drifting ice, with difficulty made about 14 miles eastward by three A. M. the sixth, when we again came to anchor. At noon we got an observation for the latitude, which proved 69° 27' 43"; longitude, by time-keeper, 168° 29'. Variation of four compasses gave the mean 17° 12' 30" east. We now sent the boat on shore to haul the seine, and caught about 300 herrings; we had tried in Wolves' and Spiral Bays, but without any success.

On the 7th, at seven A. M., an officer was dispatched with a boat round the next promontory, called Barannoi Kamen, to examine the state of the ice. At nine in the evening he re-turned, and reported that the ice was compact to the very shore, leaving no kind of passage. Captain Billings walked round the promontory, and found the ice as the officer had stated. He saw an immense number of geese on a lake, and found two mammont's tusks; one of which weighed 3 poods 17$\frac{1}{2}$ lb., or 115 lb. English

weight;

weight; the other was much smaller. Mr. Bakoff was sent with a few hands to the lake to endeavour to obtain some wild geese. They proved to be in a moulting state, and he collected in a short time 98, with which he returned to the vessel. While on shore, he saw several rein-deer, but could not shoot any.

All the 8th, we had calms and variable light airs. On the 9th, at two P. M. a moderate breeze sprung up from the north-east, which brought down upon us great quantities of floating ice, and made us seek shelter by weighing and sailing westward. At four P. M. both sun and moon were at times visible; and Captain Billings took an observation for the longitude; but, flying clouds constantly obscuring one or both, it could not be exact, although it pretty nearly agreed with our time-keeper; it proved 167° 57′ 40″, and might err, perhaps, one way or the other, a few miles. We anchored in Wolves' Bay on the 10th, at eight P. M., and remained till three A. M. on the 17th, when we again attempted to go to the north-east with a moderate north north-west breeze, keeping as near the coast as convenient. We passed immense fields of ice, which obliged us to come to anchor close in-shore, after having run about 14 miles.

On the 18th, we erected a cross on an eminence. The next day, at nine A. M., observing the sea more clear of ice, we weighed with a gentle north-west breeze, shaping our course along the coast north east. Observing a cross on shore, we sent to learn the inscription, which was only " 1762." About four P. M. we passed Barannoi Kamen, and got among pretty large detached pieces of ice, on one of which we caught a stone fox. We also saw two or three seals, and with a boat-hook caught one sleeping on the water. The weather was hazy; and, the wind freshening,

by

by ten P. M. we made 30 miles. The latter part of the time the ice increafed about us, and fome of it was eight foot above the water. Our depth was 10, 11, and 12 fathom. At eleven o'clock Captain Billings thought his fituation dangerous; he therefore tacked, and ftood back again, making a fignal for the Yafafhnoi to do the fame. At noon, we came to anchor clofe in-fhore off Barannoi Kamen, which promontory we had paffed 15 miles, being about half-way to the next point of land, called by Shalau-roff Pefofhnoi Muis, and which is the fouth-weft cape of the Tfhaoon Bay.

On Tuefday the 20th July, at fix o'clock in the morning, Cap-tain Saretfheff came on board the Pallas, in confequence of a meffage fent him; when Captain Billings informed him, that he was refolved to give up all thoughts of any further attempt, and meant to return to Neizfhni Kovima as foon as the wind would permit.

On the 21ft, at noon, we got a fight of the fun from on board: the latitude proved 69° 35′ 56″; longitude, by time-keeper, 168° 54, Barannoi Kamen bearing fouth, diftant three miles. Pefofh-noi Muis eaft, diftant 30 miles. Variation of the compafs 17° 40′ eaft.

A frefh breeze from the weft continued, with ice drifting to the eaft with the current, which now fet at the rate of three miles uniformly eaft, till midnight of the 25th July; when we obferved the current fetting in the fame direction at one mile, with little wind from the north-weft, which fhortly after veered to the north-eaft. Till this time we found the water frefh enough to drefs food, and fometimes quite frefh. With the north-eaft wind, we

obferved

obferved the current fhift to the weft, and the water became falt: we faw feveral feals, fome fmall whales called the Belluga, and one whale of a moderate fize; circumftances which induced me to think that we now might gain a paffage. Mr. Saretfheff was firmly of my opinion, and offered to attempt it in our open baidar with fix hands, meaning to fleep on fhore every night. The poffibility of fuccefs was farther confirmed by Mr. Shmaileff; but was not agreed to by Captain Billings, who took the fignatures of the officers in teftimony that it would be more prudent to return to Neizfhni Kovima. We experienced a conftant fucceffion of fnow, rain, and fogs; and the thermometer varied from the freezing point to 4° above it; nearer fhore 8° and 7°. At feven o'clock in the morning of the 26th we weighed anchor, and ftood away to the weft for the river Kovima; and, after encountering fome difficulty in getting over the flats at its eftuary, arrived on the 29th July, at eight o'clock in the morning, at Neizfhni, delivering up the veffels and ftores to the commander of the place.

To conclude the detail of this fhort excurfion, I fhall fubjoin the following remarks: The coaft of the Icy Sea is moderately high, formed by projecting promontories and fhallow bays, expofed to every wind except the fouth. The mountains are covered in different places with fnow; which melting, produces fmall torrents rufhing into the fea. They are compofed of granite, quartz, and a hard black ftone; and produce mofs; a kind of vetch, the root of which is edible; creeping willow; and birch, not exceeding ten inches in height. The fhores are covered with drift wood nearly to Barannoi Kamen, but no farther eaft. Along the fhore are numerous remains of huts, and places where fires

2

have

have been, which, in all probability, have been made and left by different hunters.

The quadrupeds that we faw were rein-deer, pretty numerous; bears, but none white; wolves, foxes, ftone fox, wild fheep, and the whiftling marmot. The birds were, gulls of feveral forts, ravens, hawks, black-headed buntings, fnow-larks, a few partridges, geefe, ducks, and divers.

The productions of the fea are very few. We frequently hauled the feine, but only once caught the feld (herring) and muk-foon (a fmall fpecies of falmon). We faw feveral belluga, feals, and one whale, but no traces of fhell-fifh of any kind. The water was frefh to a confiderable diftance; the ice we frequently tried, but found it brackifh, with neither ebb nor flow. The currents were very irregular, feldom fetting any one way longer than the wind blew, at the unfettled rates of half a mile, a mile, and three miles and a half, per hour.

The atmofphere was cold and chilly, the greateft heat that we experienced being while at anchor clofe in with the land in Wolves' Bay on the 15th July, when we had feveral claps of thunder. We had a gentle fouth-eaft breeze, and calms; and while the wind blew, the thermometer rofe to 14° and 16° above the freezing point of Reaumur. During the intervening calms, it funk to 6°, 7°, and 8°. The coldeft day was the 12th July, the thermometer being then 2° below the freezing point. It frequently indicated 1° above 0 at the time when our rigging was incrufted with ice.

The

The fogs here are very remarkable, continually hovering above the ice at no great height. At a diſtance they appear like iſlands in a haze; ſometimes like vaſt columns of ſmoke. Once, in particular, we thought that the Tſhutſki had made ſignal-fires for us; but on a nearer approach we diſcovered our miſtake.

I obſerved the horizon to be moſt clear in the coldeſt weather, and am inclined to think that this navigation ought to be undertaken about the firſt of Auguſt. The more ſucceſs is to be expected, from the teſtimony of the hunters and others who viſit theſe parts, " that the ice never breaks up until St. Elias' day, the 20th July, Old Stile (or the 31ſt July New Stile"); and I think it neceſſary to remark here, that my dates are all Old Stile, according to the cuſtom of Ruſſia.

The eſtuary of the river Kovima at Shalauroff's winter buildings, by exact reckonings of bearings, courſe, and time, from places where obſervations were taken in the Icy Sea, and from Neizſhni Oſtrog, forwards and backwards, I fix in latitude 69° 16', longitude 166° 10'; variation of the compaſs 17° 30' eaſt.

The following is the reſult of my remarks and inquiries during my ſtay at Neizſhni Kovima :—I obſerved ſwallows ſwarming together under the eaves of the church, chirping very much, particularly on the 2d Auguſt; and on the 3d there was not one to be found, nor had any body ſeen them depart. I was informed, that they made their appearance about Tzarivoi day (21ſt May), and departed on the (days of Spaſs) 2d and 6th Auguſt, never ſtaying beyond the latter date; the red-breaſt remains a day or two longer than the white. The ſnow-bunting, the firſt bird that

appears,

appears, is feen about the middle of March feeding on the feeds of grafs on the fandy fhores of the river, and about the roots of bufhes where the fun firft melts the fnow; different flights purfue each other in their migration for about a month; eagles follow clofe upon them. Swans, geefe, and ducks, arrive toward the end of April, and continue about the neighbouring lakes and rivers till the beginning of September. The river is frozen over about the 20th September, and opens about the 24th May, when it deluges the low country. The water does not retreat within its bounds till the end of June.

On the 25th November the fun fets until the 1ft January, when it again appears above the horizon; and this is the time of the fevereft cold.

M   CHAP.

CHAP. VII.

*Departure from Neizſhni Kovima Oſtrog.—Yermolova Tona.—Tow-
ing by Dogs.—The Mountain of Konzſheboi.—Seredni Oſtrog.—
Natural Hiſtory of the Kovima.—Sketch of the Inhabitants on its
Coaſt.—Information derived from an old Coſſac, and others, reſpecting the fate of Shalauroff.—Arrival at Yakutſk.*

CAPTAIN BILLINGS, Dr. Merck, Mr. Robeck, our Ruſſian
ſecretary Vaſſiley Diakonoff, and I, with a neceſſary number of
ſailors, departed from Neizſhni Kovima Oſtrog on the 6th of
Auguſt at four P. M. with the two boats and the baidar, in 4° of
froſt, rowing and hauling againſt the ſtream. At eleven o'clock
at night we arrived at Yermolova Tona, a fiſhing place reſorted
to by the inhabitants of Neizſhni during the ſeaſon. Mr. Saretſheff and the reſt of our company were left to follow us in the
tranſport veſſel, which was hourly expected to bring proviſions
for the Coſſacs. On arriving here, we were informed that this
veſſel had paſſed about two o'clock in the afternoon; but, owing
to hazy weather, we did not ſee it. We were alſo told, that difpatches from Ruſſia were in the poſſeſſion of a courier paſſenger.
A boat was immediately ſent, and we waited till the next noon
for our papers, chiefly letters. I was favoured with one from St.
Peterſburg, and another from Brigadier-General Troepolſky of
Irkutſk.

We proceeded at two P. M., having obtained one lodka, or
canoe; but found extreme difficulty in getting on with our boats

and

and fo much baggage; in confequence of which, Captain Billings left every article, except the provifions, with Vaffiley Diakonoff on fhore, in the morning of the 9th, and difpatched the baidar to Mr. Saretfheff with intelligence of the contents of our papers, &c. At fix o'clock in the evening, we arrived at three uninhabited huts belonging to the villagers of the river Omolon, and took poffeffion for the night, which was very ftormy, with fnow. The thermometer was at o; and, the gale continuing all the next day with fnow and rain, we were glad to keep fo good a birth. One of our failors, a chriftened Koriak, who formerly refided a fhort time on the Omolon, recommended, as the moft eligible method of getting forward, canoes to be drawn by dogs on the beach againft the ftream of the river. He told Captain Billings, that the village was only ten verfts by land acrofs the oppofite cape, and that he knew the road perfectly well. His advice was taken; and on the 11th, at noon, the wind abating, I received the Captain's directions to take the boats and men to the huts oppofite the difcharge of the Omolon, where he purpofed meeting me; and he, Dr. Merck, and Mr. Robeck, fet out with the failor before mentioned, and a foldier. The width of the river being about a mile and half, and the wind blowing very frefh from the weft, with great difficulty they gained the oppofite fhore, quite wet; the boat returned at fix P. M.

Bad weather detained me till the 13th in the morning, when it blew a moderate breeze, and I fet off with two boats and the canoe. After making, with great difficulty, 15 verfts, half-way to the huts, the wind increafing to a gale, I was obliged to take fhelter under the high eaftern fhore, where I paffed the night, and arrived at three o'clock in the afternoon of the 14th at the huts,

M 2
which

which Captain Billings had reached but two hours before me. We were now 110 verſts from Neizſhni.

The Captain told me, that, owing to the difficulty they experienced in croſſing the Kovima, and the ſwampy road that they had to travel, which was knee-deep in wet moſs, they did not arrive till the next noon at the habitations, after paſſing a ſhocking night, ſleeping on the moſs, in the ſnow and wind, without any covering or ſhelter : Dr. Merck and Mr. Robeck, the ſurgeon, had their toes frozen.

Sunday the 15th Auguſt, at eight o'clock in the morning, Captain Billings, myſelf, two attendants and four guides from the Omolon village, proceeded on our voyage to Seredni, leaving Dr. Merck and Mr. Robeck to follow in the boats. Croſſing the river, we put our harneſſed dogs on the beach, and they hauled us 40 verſts, to a famous mountain called Konzſheboi, where we pitched our tent, and paſſed the night. We obſerved wild onions, thyme, tanzy, tſhornoi golovnik, currant and roſe buſhes, about the ſhores ; and in the fiſſures of the rock, juniper, creeping cedar, and here and there a ſtunted larch-tree. The compoſition of the mountain is granite and quartz. On the beach were numberleſs ſmall pebbles of carnelian and calcedony.

We conſtantly travelled at the rate of 50 or 60 verſts each day, until the 22d, when we arrived ſafe at Seredni Oſtrog, 460 verſts from Neizſhni.

The eaſtern ſhores of the river are uniformly mountainous, producing agates, jaſper, porphyry, and cryſtals ; and we ſaw ſeveral Yakuti on hunting parties.

On

On the 25th, Dr. Merck and Mr. Robeck arrived; and Mr. Saretſheff, with all our party, joined us in the tranſport bark on the 28th.

Of the neighbouring Yakuti we ſent to requeſt a ſupply of horſes to convey us to Yakutſk, with the earlieſt winter roads. The inhabitants were, for the greater part, at their autumnal fiſhing huts on the river Euxeva, 40 verſts up the Kovima, when we arrived; but returned a few days after with a plentiful ſupply of fiſh, berries, roots, &c.

On the 20th September the river was frozen over; and on the 22d, the inhabitants made a kind of dam nearly acroſs, by ſticking poles upright quite cloſe together, only leaving openings for inſerting nets and wiers, to catch a winter's ſupply of freſh fiſh.

The nets were examined twice a day, and generally found well ſtocked with nelma, mukſoon, omul, and ſeld (a kind of herring), a few ſtirled alſo were now and then caught; and theſe were thrown on the ice to freeze, the only method adopted to preſerve them. The weather was clear and cold, with from 5 to 10 and 16° of froſt.

The river Kovima takes its riſe from the Virchoyanſky chain of mountains, and flows in a direction nearly north-eaſt about 1800 verſts. Virchni, or the upper Oſtrog, is about the middle of its courſe, and very few fiſh aſcend higher. Near its ſource are three huts and a ſtore-houſe, called Virſhinoi, where government ſupplies of proviſions are houſed, and barks built for their conveyance down the river.

Fiſh

Fish are very plentiful, of which the following sorts are caught:

English Names.	Yakut Names.	Russian Names.
Sturgeon.	Katus.	Osètre; called by the inhabitants of the Kovima Shtshalbysh—July to October.
Salmon.	Tut Balyk.	Nelma; large white salmon 2½ to 4 feet long, weighing upwards of 60 pounds—July to October.
Ditto.	Mungur.	Chir or Tshir; 20 inches—May to November.
Ditto.	Muksoon.	Muksoon; 15 to 18 inches, silvery scales—Sept.
Ditto.	Omul.	Omul; 12 to 14 inches—Ditto.
	Seld.	Seld; resembling a herring, silvery loose scales—Ditto.
	Shookur.	Sieg; 12 to 15 inches, silvery scales—May to November.
Quab.	Selu Sar.	Nalime; to 5 feet long, resembling the cod in shape and taste; has one beard; the liver extremely large; and I have extracted a full pint of fine pellucid oil from one: by putting it over a slow fire in a frying-pan, and cutting it, almost all the liver has been dissolved.
	Baring Ata.	Peledi; somewhat resembling a carp in shape, but quite white scales: bony.
	Tshukutshan	Tshukutshan; about 20 inches long, quite round and firm; about 7 inches in circumference in the thickest part, gradually tapering to the tail, which is forked. It has two dorsal fins, very compact and fine scales, and a thick skin. The head rather flat; pointed gristly nose; the mouth is underneath, about 2 inches from the tip, shaped nearly like that of the leech, without teeth; a very bony fish; flesh white. They are rather scarce, and not esteemed for food.
	Booyit.	Lenok.
		Koniok.
	Dyrga.	Charius.
	Kiustak.	Chebak.

Perch.

Englifh Names.	Yakut Names.	Ruffian Names.
Perch.	Alfhre.	Okun, } plenty in all the ftony rivulets.
Stone ditto.	Taafbas.	Yerfh ; }
Char.	Sobo.	Karas ; chiefly in lakes.
Trout.	Kafil balik.	Krafnaia riba ; fcarce.
	Irungk bulyk.	Nefnaki ; a white fifh fhaped like a trout ; very fcarce.
	Timir atta.	Zfheleznoi noga.
	Turuchan.	Nerpifki ; fhorter and broader than the feld.
Minnow.	Soluro.	Mondufhka.
Pike.	Sording.	Shtfhuk ; fome of a moft extraordinary fize. I faw one caught in the lake Kyfla, near Zafhiverfk, about 6 feet long, and weighing 108 lb. or 3 poods. The back, towards the head, was covered with a kind of mofs. I ate fome of it, and found very little difference between the flavour of it and the fmaller fort.
	Irungka, - -	refembling in fize and appearance a fprat or anchovy ; perhaps the fame kind of fifh as is caught at Revel, and called Strömlingi.

The mukfoon, omul, and feld, come in very great fhoals in September, are very numerous for about 10 or 15 days, and depart fhortly after the river clofes. They do not afcend fo high as Virchni. Nelm, tfhir, or chir, and fieg, are caught all the year as high up as Virchni ; and the greater part of fuch as are caught in the fpring and fummer are fplit and dried, and the bones taken out, from which the inhabitants extract a great quantity of oil, as alfo from the fat about the guts. The feld yields a great deal. What I have called the fturgeon is, in my own opinion, the ftirled. I do not know the difference (nor did any one in our Expedition), but judge from the fize ; for 1 never faw one that weighed more than 40 lb., and the ordinary weight was from 5 to 10 lb. ; yet, owing to their extreme fatnefs and firmnefs, I am inclined to think that the rivers of thefe parts are favourable to them ; and the fturgeon that I have feen in other

parts

parts are three times as large, without being fo fat, firm, or well flavoured : all, except this one fort, are caught as well in lakes as rivers, where they pafs in the floods and thrive very well, particularly the tfhir and fieg.

The inhabitants fifh with the feine in fummer ; and in winter they plant ofiers nearly acrofs the river, fo clofe as to preclude the fifh from paffing ; but leaving openings for wiers and nets.

The Beafts that infeft the neighbourhood are :

Englifh Names.	Ruffian Names.	Yakut Names.	Time of the Chafe.
Elk.	{ Sochata. { Lofs.	Toyak.	} September, October, and November.
Deer.	Olen.	Miniak.	

Spring bows are fet, with a ftring leading to the path which they take. In April, and the beginning of May, they are chafed on fnow-fhoes. At this feafon, the fun in the day-time thaws the fnow, which the night froft hardens enough to fupport a man and dogs, though the animals always break through, and cannot extricate themfelves. Immenfe numbers of deer are flain in Auguft, on fwimming acrofs rivers in returning to the woods from the borders of the Icy Sea, whither they retreat in fpring to fhelter themfelves from the flies and infects that infeft the forefts. Their migration is very curious. They herd all together ; and I am told, that the males form the van and rear, while the females are inclofed in the centre : Bears and wolves follow them, deftroying fuch as ftraggle from the main body ; foxes lag behind and clear the remains. Thus do I account for the appearance of bears,

wolves,

wolves, and foxes, fo far beyond the forefts. Eagles and other birds of prey hover over the deer at the time of their migration, and give the earlieft intimation to the hunters of their near approach. On their taking the rivers, the hunters man their canoes; two men with fpears in fome, while others are occupied by boys and women, furnifhed with long cords, which they throw over the horns of the ftabbed deer, and tie one end to ftakes or trees on fhore. I have not, however, been fortunate enough to fee them at this time—Price, elk fkin 2, doe fkin, 1 ruble.

Englifh Names.	Ruffian Names.	Yakut Names.
Bear.	Medved.	Ehea; from May till late in September—Middling Effe; fkin, 1 ruble.
White Bear.	Beloi Medved.	- - - ; about the Icy Sea. We did not fee any—1 ruble.
Glutton.	Ryfomag.	Siégan and Begó ; all the winter; not numerous—2 to 10 rubles.
Wolf.	Volk.	Beréh ; never fought after in thefe parts—2 to 8 rubles.
Fox.	Leefits.	Safil ; numerous, and much looked for in October and November—According to quality, 1 to 5 rub.
Stone Fox.	Peffets.	Kirfa ; October and all the winter—50 copeaks.
Ermine.	Gornaftal.	Belilak ; all the winter in woods near habitations, and frequently about the flour magazines—5 copeaks.
Lynx.	Rys.	E-us ; autumn and all winter—3 to 10 rubles, according to the length of hair.
Otter.	Vüidra.	Itie ; fummer—8 and 10 rubles.
Sable.	Sobol.	Kies ; very feldom caught about the Kovima—10 rub.
Sheep. Argali.	Baran.	Tfhubek ; about the mountains at the fource of the Kovima, and all over the Virchoyanfki chain to Kamtfhatka—1 ruble.
Hare.	Zaits. Ufhkan.	Kobach ; all the winter; but chiefly when the firft fnow falls—3 to 5 copeaks.
Marmot.	Tarbagan.	Tarbagan, Kutier ; much efteemed by the Yakut for food and drefs. They pafs the winter under
Ditto.	Suflik.	ground,

N

ground, have several chambers in their holes, and lay in a considerable stock of grass and sweet roots; also the nut of the cedar—5 to 10 copeaks.

English Names.	Russian Names.	Yakut Names.
Squirrel.	Belka.	Tee-ing; spring and autumn—3 to 5 copeaks; esteemed good eating by the different tribes.
Flying do.	Letushka.	Tirik-annat; *annat* is winged; *tirik* is skin—Of no value.
Striped do.	Burunduk.	Burunduk; 2 copeaks.

The three species of squirrel pass the winter in hollow trees, or under ground, in the same manner as the marmot.

English Names.	Russian Names.	Yakut Names.	
Mountain. Whistling rat.	Pishuka.	Kyla.	never sought after—Worth nothing.
Sharp nosed. Com. mouse.	Müish.	Kutuyak;	

The Birds consist of the following kinds :

English Names.	Russian Names.	Yakut Names.
Swan.	Lebed.	Kubah; appear about the 20th April, and depart in September.
Goose.	Goose.	Kaas; appears and goes a few days later.
Duck.	Utka.	Kus; are in great variety, and numerous.
Eagle.	Orel.	Baruldo; black.
Ditto.	Ditto.	Toyon; white head and tail.
	Skopa.	Umsan; of the eagle or hawk kind: darts in the water for fish.
	Yastrip.	Kirt.
	Kretchet.	Ditto.
	Korshoon.	Togolak.
	Sokol.	
Owl.	Filen.	Moksoghol and Karali.
Gull.	Chaika.	Kopta.
Small black head. Gull.	Marteshka.	Tiraghi.
Crane.	Zshurav.	Turuja.
Stork.	Sterch.	Kutelik.
Partridge.	Kuropatka.	Kabdshi.
Raven.	Voron.	Sor.
Crows.	Varonna.	Tarak.

Divers.

Englifh Names.	Ruffian Names.	Yakut Names.
Divers.	Gagara.	Koghas.
Black Game.	Tetere	Ulöer.
Black woodpecker	Dfholna.	Kirgil.
Woodpecker.	Datel.	Tonoghas.
Swallow.	Laftofhka.	Karangachuk.
Thrufh.	Drofd.	Tatfheger.
Snow-bunting.	Snegir.	Tulak.
Snipe.	Kulik.	Sulbaraga.
Cuckoo.	Kokufhka.	Kuga.

The eagle and hawk kind, I am informed by the Yakuti, as alfo by the different inhabitants here, are dormant in hollow trees during the winter.

Lift of Trees, Bufhes, and Berries.

Larch—This is the chief tree in ufe, for building, firing, &c. and the moft plentiful. It is pretty fizeable as far as Virchni, and the country is moderately wooded about 200 verfts lower, but the trees very ftunted: beyond that, they are in clufters on elevated fpots of ground to about 30 verfts from the Icy Sea, where they ceafe growing, in about the latitude of 68° 30´.

Birch; extends to a little below Seredni; but very ftunted and fmall trees.

Poplar and Afp; grow to a moderate fize on the iflands fheltered by mountains, about the fource of the Kovima; but do not extend fo low down as Virchni.

Mountain Afh; plenty as far as Virchnoi, but very fcarce lower down.

Alder and Willow; have a trunk about 18 inches in circumference, and grow to the height of 2 fathom about Virchni. They gradually diminifh in fize, and ceafe growing with the larch.

Creeping cedar, brufhwood, black and red currant, rofe and juniper, are met with as low as Neizfhni. Brufhwood and creeping willows extend to the Icy Sea, but never exceed from 6 to 8 inches. The creeping cedar, or pinus cembra, produces a confiderable quantity of feeds or nuts in cones, like the common pine; but they ripen only the fecond year. Immenfe numbers are collected by the inhabitants; fometimes a confiderable quantity are found in the fquirrels' nefts in hollow trees; in fact, they are the chief food of fquirrels and mice. A very pellucid and fweet oil is extracted from thefe feeds.

Berries.

Mountain afh berries; are gathered, and ufed to give a pleafant flavour to their drink.

Black and red currants; collected in abundance, and preferved in cafks among ice; fome

are

are boiled and preferved.   The black only extend to about Seredni; but the red continue growing as far as Neizſhni.

Cranberry – Theſe are ſcarce, and extend no farther north than Seredni : they are always
preferved raw.

Bruſniki; Vaccinium vitis idæa ; Whortleberry—Theſe are very plenty as far as Neizſhni,
and are preferved raw.

Golubniki; are very numerous.   They ſeem to. delight in ſuch ſtony places .as are
overflowed in the ſpring.   They are very. pleaſant-taſted ; of a dark blue colour;
and grow on a low buſh exactly reſembling a myrtle.   They are preferved by. boiling.

Maroſhka ; Rubus chamæmorus—Theſe are the favourite berry of the inhabitants, and
grow in damp moſſy places, particularly near lakes.   They are reckoned a certain cure
for the ſcurvy ; and are always preferved raw.

Siecha ; growing on dry ſtony places about the mountains, on a creeping ſpecies of heath,
with ſhort needle leaves ; they are very ſmall, black, and ſtony, are collected in great.
abundance, and preferved by boiling.

Knezſhnitſi ; Rubus Arcticus ; are ſcarce, growing about the roots of the alder and currant buſhes.

The inhabitants of theſe parts prepare their food in the following manner :

Beſides boiling and frying fiſh, as is done in every country,
ſoups are made of quabs, karas, and perch.   The upper part of
the head or griſtle of the nelm, ſieg, and tſhir, are boiled, and
ſerved up cold as a whet, with ſalted onions, and the juice of
cranberries inſtead of vinegar.   They bone boiled fiſh, then beat
them in a mortar to the conſiſtence of paſte, make it into the
form of a pie, putting into it either the ſeld, the heads of ſalmon,
or, which is reckoned beſt, the liver of the quab, and bake it, with
or without onions.

The ſpawn of fiſh beat up in a mortar, ſometimes mixed with
flour, and fried with onions, is called baraban ; if fried like a
                                                        cake

cake without onions, and preferved berries put on the top, it is called fhangee.

Pike are fkinned, and beat up raw, with onions, wild thyme, and pepper; made into force-meat balls, and inferted in foups and fifh pies ; and fometimes made into cakes, and fried. They are called'telnée.

The thick gut of fifh, particularly the quab, is boiled, and ferved up cold, with different berries, by way of defert after dinner.

Their drink is the fermented juice of berries mixed with water. They make vinegar, or rather a good fubftitute for it, by fermenting onions with flour, or the pounded inner bark of the larch ; and I thought it very good.

An infufion of wild thyme, of dog-rofe leaves and ftalks, and of the plant called tfhernoi golovnik, is ufed inftead of tea.

Tufks of the mammont are found very numeroufly about the fandy high fhores of the river, at a confiderable depth ; and the fpring floods wafhing away the fand difcover them. I am not at all furprifed at their being buried fo deep ; for every fpring the flood leaves immenfe quantities of fand and earth on the fhores of the rivers ; perhaps to the depth of two to three inches, and among bufhes much more. They are equal to elephants' teeth in whitenefs and beauty, but very different in their fhape, being all bent fpirally, forming about one round and a half. The largeft that we found, which was on the fhores of the Icy Sea, meafured as follows, French meafurement :

Length,

	Feet.	Inches.	Lines.
Length, with the bend, - -	8	7	4
Diſtance from one end to the other, ſtraight, -	4	1	9
Circumference near the root - -	0	14	3
The thickeſt part 22 inches from the root -	0	17	8
Of the middle - - -	0	15	8
Of the point - - -	0	9	5

Weight $137\frac{1}{2}$ lb. Ruſſian weight, equal to 115 lb. weight avoir-dupois.

The outſide was very brown from its having been expoſed to the weather; and it was cracked through the coat, or upper ſtratum, about an inch. The inſide was quite firm, and very white.

The horns of another animal are frequently found, adhering to a part of the ſkull, and reſemble very much thoſe of the buffalo. The elaſtic part of theſe are much eſteemed by the Tungooſe, &c. for ſtrengthening their bows.

I am ſorry that my want of knowledge in natural hiſtory, mi-neralogy, and botany, prevents my giving a better account of theſe almoſt unknown parts. Had we been accompanied by any ſkilful perſon, I ſhould have made this ſcience a chief part of my ſtudy. Situated as I was, I obſerved every circumſtance as well as I could, and communicate my remarks in the beſt manner I am able.

I ſhall now lay before my readers the reſult of my inquiries among the inhabitants.

Daniel

Daniel Tretiakoff, a Coſſac in the 90th year of his age, gave me the following intelligence :—" I came here in 1739 with a " commiſſary, who was ſent to collect tribute ; and I was de- " tained here as interpreter by Laptieff, who made an attempt to " croſs the Icy Sea, and returned late in the ſame autumn. Virch- " ni was then inhabited by exiles, who were trading pedlars. " Yukagers were very numerous then ; and I believe they de- " rived the name from one of their warriors : thoſe of the Omo- " lon, were called Tſheltiere ; thoſe of the Alaſey, Onioki ; and " thoſe of the Anadyr and Annui, Tſhuvantſi and Kudinſi. Wars " with the Tſhutſki and Koriaks, and fatal diſeaſes, have almoſt " extirpated the race. I have heard of a numerous nation inha- " biting the Kovima, called Konghini, and think it was from " them that the river obtained the name of Kovima. Remains " of many villages were ſeen on the borders of the river, and " numbers of ſtone hatchets, and ſtone pointed arrows, have been " found about their ruins.

" There were but very few Yakuti when I firſt came ; and I " believe that none of them were here 70 years back. The pro- " viſions for ſupplies to Kamtſhatka and Anadirſk uſed to be ſent " from the Kovima, up the river Annui, and down the Anadir. " At that time traders frequently viſited us, and very fine ſables " were caught in abundance, particularly about the Omolon.

" On Pavlutſki's return from his firſt attempt to ſubdue the " Tſhutſki, the oſtrog at Neizſhni was full of women priſoners. " Numbers were returned ; ſome he attempted to ſend to Ruſſia, " but every one of them died on the road."

He gave me the following account of Shalauroff's expedition in 1762 :   " In

" In the beginning of the year, Ivan Bachoff, his affociate, an
" exiled naval officer, died at Neizfhni, and left Shalauroff to exe-
" cute the enterprife alone.   About St. Elias's day he weighed
" anchor from his winter buildings at the eftuary of the Kovima.
" His followers were exiles and runaway foldiers, not hired to
" receive pay, but volunteers, to receive a proportionate fhare of
" the produce of the voyage, intended in queft of ivory and furs.
" Of fuch as could write and read he made officers, and the
" fubordinates were mutually agreed upon.

" He had failed but a very little way before he encountered
" contrary winds, which detained him till the 10th of Auguft.
" Much ice was in fight, but none near the veffel.   He now
" kept well in with the fhore, paffed Barannoi Kamen, and
" reached a point of land to the eaft, which may be feen in clear
" weather.   Here the ice inclofed them three days, and damaged
" the rudder, which, however, was foon repaired.   This point
" of land is the fouthern cape of a deep bay, at the entrance of
" which is an ifland of moderate fize.

" The weather was very cold, and the crew wanted to feek for
" a wintering place.   Shalauroff, finding the fea moderately clear
" of ice, endeavoured to perfuade them to go farther; to which,
" however, they would not agree; and on the 25th Auguft he
" fteered into the bay, round the northern extremity of the ifland,
" to feek a place wherein they might pafs the winter; but as there
" was neither wood nor fifh to be obtained, and his crew would
" not liften to his perfuafions to continue their voyage, he was
" compelled, againft his inclination, to return to Neizfhni.   Here
" his companions difperfed, but he himfelf went to Mofco.   In
" 1764 he undertook another voyage under the fanction of govern-
" ment; but he never returned, nor was afterwards heard of."

Dauerkin,

Dauerkin, our Tſhutſki interpreter, aſſured us, that Shalauroff's veſſel was found drifting, near the mouth of the Kovima, in the autumn of the ſame year that he put to ſea ; and that his people were found frozen to death about 20 or 30 verſts eaſt of Barannoi Kamen in a tent, with proviſion, ammunition, and arms. I note this piece of information, although I think it very inconſiſtent, and do not believe it.

Affanaſſy Kaſſimoff, an inhabitant of Neizſhni, who formerly reſided at Anadirſk, aſſured me, that in the year 1766, or 67, the Tſhutſki brought him ſeveral paintings of Ruſſian ſaints ; that ſome of them had cloth jackets, and that they were deſirous to get gunpowder. They ſaid, that they had found theſe things on ſhore. This was in the ſpring of the year ; and he ſuppoſed that it was the property of Shalauroff and his people, of which they had been pillaged, and afterwards probably murdered, by the Tſhutſki. They ſaid that the articles were found to the north of the bay of Anadyr. I am inclined to think that Shalauroff doubled the capes, and was cut off in attempting to paſs the winter among the Tſhutſki.

On the 23d September we obtained a few horſes, and on the 25th diſpatched the firſt party to Yakutſk with Mr. Bakoff. On the 28th, Mr. Saretſheff departed with the chief hands. Captain Billings and I followed on the 8th of October. We croſſed the Alaſey mountains at the ſource of the river of that name, and came into the ſame road that I had taken from Zaſhiverſk, at which place we arrived the 22d October, and remained there three days : we then proceeded, croſſed the Virchoyanſki chain at the ſource of the Yana, and arrived at Yakutſk on the 13th November, after ſuffering inconceivable hardſhips from the ſeve-

O   rity

rity of the cold, and travelling on horfeback. I computed the distance at 1300 verfts in the fummer feafon ; but have every reafon to believe, that it muft be 2000 or 2300 verfts, when travellers are compelled to go round the bogs and lakes, and to feek fordable places in the rivers : the Yakuti and Ruffians call it 2500 verfts.

On the fouth fide of the Virchoyanfki mountains, the face of the country is lefs barren ; and, in addition to the trees before mentioned, are the fir and common pine in abundance, and of large fize.

CHAP. VII.

Meet with Mr. Ledyard, who travels with the Command to Irkutſk.—
He is arreſted by an order from the Empreſs, and ſent under a guard
to Moſco.—The Governor-General, Jakobi, called to St. Peterſburg.
—The Command arrives at Yakutſk.—Some particulars reſpecting
Laſhoff's Travels to the Icy Sea, 1770-3.—Chvoinoff's Journey
thither in 1775.—The Command arrives at Ochotſk, but returns
immediately to winter at Yakutſk.

AT Yakutſk we found, to our great ſurprife, Mr. Ledyard, an
old companion of Captain Billings, in Cook's voyage round the
world; he then ſerved in the capacity of a corporal, but now
called himſelf an American colonel, and wiſhed to croſs over to
the American Continent with our Expedition, for the purpoſe of
exploring it on foot.

Captain-Lieutenant Bering, who had been ſent the 12th of
February laſt from the Kovima, to ſuperintend the forwarding
the neceſſaries for the Expedition to Ochotſk, was alſo here. He
had forwarded many articles during the ſummer, and ſent ſome
of the anchors and heavy baggage to the river Mayo, to be tranſported to Yudomſki Kreſt by the water communication. The
guns, medicines, ſailors' clothing, &c. weighing upwards of 100
tons, ſtill remained at Irkutſk, where they had lain ever ſince laſt
winter.

Captain Billings reſolved to go himſelf to Irkutſk to ſee theſe
articles forwarded down the Lena ſo ſoon as the river ſhould open in

O 2   the

the fpring. Accordingly, on the 29th December, he fet out with carriages on fledges, which we had made on purpofe. Mr. Ledyard, Robeck, Leman, his firft mate, and I, accompanied him; the Ruffian fecretary and feveral neceffary hands were ordered to follow with all poffible fpeed.

We arrived the 16th January 1788, and I took up my abode with my friend Brigadier Troepolfki.

The Captain began making preparation for tranfporting the guns, &c. and fent to build veffels on the Lena at Katfhuga, where they were depofited.

In the evening of the 24th February, while I was playing at cards with the Brigadier and fome company of his, a fecretary belonging to one of the courts of juftice came in, and told us, with great concern, that the Governor-General had received pofitive orders from the Emprefs, immediately to fend one of the Expedition, an Englifhman, under guard to the private inquifition at Mofco; but that he did not know the name of the perfon, and that Captain Billings was with a private party at the Governor-General's. Now, as Ledyard and I were the only Englifhmen here, I could not help fmiling at the news, when two huffars came into the room, and told me that the Commandant wifhed to fee me immediately. The confternation into which the vifitors were thrown is not to be defcribed. I affured them that it muft be a miftake, and went with the guards to the Commandant. Here I found Mr. Ledyard under arreft. He told me, that he had fent for Captain Billings, but he would not come to him. He then began to explain his fituation, and faid that he was taken up as a French fpy, whereas Captain Billings could prove the contrary;

trary; but he fuppofed that he knew nothing of the matter, and requefted that I would inform him. I did fo; but the Captain affured me that it was an abfolute order from the Emprefs, and he could not help him. He, however, fent him a few rubles, and gave him a peliffe; and I procured him his linen quite wet from the wafh-tub. Ledyard took a friendly leave of me, defired his remembrance to his friends, and with aftonifhing compofure leaped into the kibitka, and drove off, with two guards, one on each fide. I wifhed to travel with him a little way, but was not permitted. I therefore returned to my company, and explained the matter to them; but, though this eafed their minds with regard to my fate, it did not reftore their harmony. Ledyard's behaviour, however, had been haughty, and not at all condefcending, which certainly made him enemies.

I found a confiderable alteration in this city; it, indeed, ftill continued the fame hofpitable and agreeable place for a vifitor, but the harmony of the inhabitants was not fo complete. Not to tire my readers with particulars, I fhall only acquaint them, that there now exifted a difference of opinion in the town, which led to the formation of two parties. However, at the latter end of March, the Governor-General, Jakobi, a good and worthy man, who had been particularly kind to our Expedition, was called to St. Peterfburg. The heads of one party accompanied him, and harmony was again reftored.

We remained here, enjoying excellent company and good living, with every rational entertainment, till the 10th of May, when we took our departure for Katfhuga.

Thirteen

Thirteen veffels were nearly ready for tranfporting our guns, medicines, glafs, failors' clothing, and our own ftores; and on the 15th, nine veffels being completely loaded, I received the charge of their conveyance to Yakutfk. The crews confifted of 50 exiles of the worft clafs, and fix foldiers. On the very firft day, I was under the neceffity of inflicting punifhment on one of them for a theft, and forbade all perfons from leaving their vef-fel, appointing at the fame time a foldier to go on fhore for them every morning to make purchafes of provifions for the day, and allowed each man a daily portion of brandy out of my own pri-vate ftock. Whether this had any effect upon them as an indul-gence, or that my determined manner of proceeding, and the feverity of the punifhment that I inflicted, more prevailed, I can-not tell; but I never faw people more active, attentive, and obe-dient, than they were all the reft of the way.

I arrived fafely on the 4th of June at Yakutfk, and immediate-ly croffed over with all the veffels to the oppofite plains. On the 6th, I difpatched 150 horfes, properly loaden, for Ochotfk, under the charge of fome foldiers. Horfes were kept in readinefs by Captain Saretfheff and Mr. Bakoff. The former gentleman took charge of the guns and all heavy materials, and conveyed them acrofs the country about 300 verfts to the river Mayo, where he had prepared veflels for their conveyance againft the ftream of that river and the Yudoma. On the 8th, Captain Billings arrived with the remainder of the articles, the greater part of which were ftill unpacked, particularly cloth, yarns, &c.

By the 15th of July, every article was forwarded, and all our hands, except a few attendants, and our naturalift, Dr. Merck, who went

7

early

early in the fpring to the neighbourhood of the Viluye, or Vilui, to obferve and collect the productions of thofe parts, from which excurfion he did not return till the beginning of Auguft.

During my ftay in Yakutfk, I made it my particular bufinefs to get acquainted with Lachoff and his companions, with a view of obtaining fome information concerning his travels to the Icy Sea. Lachoff was old and infirm, and recommended me, for any intell'gence that I required, to one of his companions, Zaitai Protodiakonoff, now a burgher and fhopkeeper in this town.

Protodiakonoff accompanied Lachoff in 1770 from his winter buildings at the eftuary of the Yana, in the month of March, to Swatoi Nofs, the northern promontory of a bay which receives this river.

They faw an immenfe herd of deer going to the fouth, and obferved that their traces were from the north acrofs the Icy Sea. Lachoff refolved, if poffible, to find out whence they came, and in the beginning of April fet out very early in the morning, with his nart drawn by dogs. Towards evening he arrived at an ifland, 70 verfts from the promontory, in a due north direction, where he paffed the night, and the next day proceeded farther, the traces of the deer ferving as a guide. About noon he arrived at a fecond ifland, 20 verfts diftant, and in the fame direction. The traces coming ftill farther from the north, he continued his route. At a fmall diftance from the fecond ifland, he found the ice fo rugged and mountainous, as to prevent his proceeding with dogs. He obferved no land ; and therefore, after paffing the night on the ice, he returned, and with great difficulty, for want of provifions for his dogs, regained Swatoi Nofs. He reprefented his

discovery

difcovery to the Chancery of Yakutfk, and the intelligence was forwarded to St. Peterfburg. The Emprefs Catherine II. called the iflands by the name of the difcoverer, and gave him the exclufive right of collecting ivory and hunting animals in this place, and in any other that he might thereafter difcover.

In 1773, he went with five workmen in a boat to the iflands, and continued acrofs ftraits, where he found the fea very falt, and a current fetting to the weft. He foon faw land to the north, the weather being pretty clear, and arrived on what he called the third ifland. The fhore was covered with drift wood. The land was very mountainous, and feemingly of great extent; but no wood was feen growing, nor did he obferve the traces of any human being. He found fome tufks of the mammont, faw the tracks of animals, and returned (without making any other difcovery) to the firft ifland, where Lachoff built a hut of the drift wood, and paffed the winter. One of his companions left a kettle and a palma on the third ifland.

This was reckoned a difcovery of fome importance, and the land-furveyor Chvoinoff received orders from the Chancery of Yakutfk to accompany Lachoff to this fartheft land, and take an exact furvey of the fame. In 1775, on the 9th February, he left Yakutfk, arrived on the 26th March at Uft Yanfk Zemovia, or winter huts, at the eftuary of the Yana. He immediately proceeded acrofs the bay to Swatoi Nofs, which is 400 verfts from the difcharge of the river, in a direction north north-eaft. On the 6th May he arrived at the firft ifland, which is 150 verfts long, and 80 verfts broad, on the wideft part, and 20 verfts on the narroweft. In the middle is a lake of confiderable extent, but very fhallow, and the borders of which are fteep. The whole ifland,

except

except three or four inconfiderable rocky mountains, is compofed of ice and fand; and, as the fhores fall, from the heat of the fun's thawing them, the tufks and bones of the mammont are found in great abundance. To ufe Chvoinoff's own expreffion, the ifland is formed of the bones of this extraordinary animal, mixed with the horns and heads of the buffalo, or fomething like it, and fome horns of the rhinoceros; now and then, but very rarely, they find a thin bone, very ftraight, of confiderable length, and formed like a fcrew.

The fecond ifland is 20 verfts diftant from this; low, and without drift wood; 50 verfts in length, and from 20 to 30 verfts broad. Here alfo the tufks and other bones are found; and great numbers of the arctic foxes are to be met with on both. The furface is a bed of mofs of confiderable thicknefs, producing a few low plants and flowers, fuch as grow about the borders of the Icy Sea. This mofs may be ftripped off as you would take a carpet from a floor, and the earth underneath appears like clear ice, and never thaws: thefe fpots are called Kaltufæ.

The ftraights to the third ifland are 100 verfts acrofs. He travelled along the fhore; and on the 21ft May difcovered a very confiderable river, near which he found the kettle, palma, and fome cut wood, in the fame place and fituation as they had been left by Lachoff's companions three years before Chvoinoff's arrival. This river he called Tzarevaia Reka, in confequence of having difcovered it on the 21ft of May. The fhores were covered with drift wood, all of it extremely fhattered. Afcending to the top of a very lofty mountain, he faw a mountainous land as far as his eye could trace in clear weather, extending eaft, weft, and north. Continuing his route along the coaft 100 verfts, he

P   obferved

obſerved three rivers, each of which brought down a great quantity of wood, and abounded in fiſh ; and here the nerk, a ſpecies
of ſalmon frequenting Ochotſk and Kamtſhatka, was in abundance, though not found in the Kovima or Indigirka.   On this
land he paſſed the ſummer, and returned in the autumn to Swatoi Noſs.

I aſked, whether he obſerved any regular ebb or flow of the
tide ? He ſaid, that " he did not obſerve any remarkable alteration."   Whether he recollected how the current ſet ? " He be-
lieved to the weſt."   Whether the water was ſalt ? " Yes, and
very bitter."   He further obſerved, that there were whales and
belluga, white bears, wolves, and rein-deer.   No growing wood
was to be ſeen, and the mountains were bare ſtone.   None of
theſe travellers took any notice of the depth of the water, nor
were they acquainted with the nature of tides.

This was the total ſum of intelligence that I was able to obtain
concerning this land ; and I am told, that ſince Chvoinoff no
traveller has paid a viſit to it.   Perhaps the three rivers obſerved
are only ſo many diſcharges running from one that is very conſiderable.

On the 11th of Auguſt we again ſet out from Yakutſk for
Ochotſk, accompanied by the Captain of the diſtrict, to examine
into the ſtate of numerous articles that had been ſcattered on the
road, owing to the loſs of horſes.   On the 23d we croſſed the
White River, without the leaſt difficulty, and arrived on the 31ſt
at Yudomſky Kreſt ; where we found our guns and heavy baggage all ſafely arrived, and were informed that Mr. Saretſheff
had ſet out for Ochotſk four days before.   I believe this to be
                                                                    the

the firſt inſtance of baggage of any kind having been tranſported from Irkutſk to this place in one ſeaſon by the water conveyance.

We arrived at Ochotſk the 6th of September, and found every thing going on in the beſt order with ſpirit and alacrity. Obſerving, however, that our ſhips could not be ready for ſea before next July, Captain Billings reſolved upon returning to Yakutſk to paſs the winter. Captain Saretſheff propoſed ſurveying the coaſt of the ſea of Ochotſk, as far as the Chineſe frontiers, in an open boat; and Captain Billings promiſed to meet him in the enſuing month of June at the diſcharge of the Aldima, to which place he purpoſed going by land and water, with Tungooſe guides from Yakutſk. Matters being thus ſettled, on the 12th September Captain Billings, Mr. Robeck, and I, again expoſed ourſelves to the dangers and difficulties of a journey on horſeback of 1200 verſts at ſo late a ſeaſon of the year. We got into ſevere winter at Yudomſky Kreſt on Wedneſday the 20th September. The next morning we had 20° below the freezing point of Reaumur, and the river was full of drifting ice; notwithſtanding which, Captain Billings attempted to go by water to Uſt Mayo; but the ſecond day we were frozen up, and obliged to return on foot to the Kreſt. We obtained horſes, and proceeded on the 27th September; but the ſeverity of the weather and bad roads prevented our reaching Yakutſk till the beginning of November 1788.

I now obſerved, that the officers of government at Yakutſk were ſuddenly become wealthy; that ſome, who with difficulty procured the common neceſſaries of life on our firſt arrival in this town two years ago, were now enabled to keep a carriage, with every thing ſuitable to that ſtyle of living; and, upon the ſtricteſt

inquiry,

inquiry, I found, that thefe gentlemen were the volunteers who were fo active in procuring horfes for the ufe of the Expedition.

During the winter, I employed myfelf in procuring the beft intelligence that I could obtain, in addition to what I already knew, concerning the Yakuti; and the refult I fhall communicate in the following Chapter.

CHAP.

CHAP. X.

Account of the Yakuti, collected from personal inquiry and research.

THE nation known among the Ruffians by the name of Yakuti call themfelves Socha, and fay that they came originally from the fouth. A nation of Mongals inhabit the diftrict of Krafnoyarfk, extending to China, who alfo call themfelves Socha, and fpeak the fame language as the Yakuti. Thefe relate the following ftory of their migration.

The Toyon (i. e. Chief) Omogai Bey, with all his tribe and cattle, left the fertile plains fituated to the weft of the lake Baical, or Baighal, to make way for a more powerful horde; retreating to the graffy meadows between Irkutfk and the river Lena, now known by the name of the Buratfki Step. Here he refided fome years, probably at continual ftrife with the Burati; for he was compelled to fly from their fury, availing himfelf of the decreafing moon, at which time the Burati never attack their enemies. Omogai croffed the Lena, at a fpot between where Katfhuga and Vercholenfk now ftands. He kept clofe to the river, making refting-places where he found pafture for his cattle, until he atrived at the eftuary of the Olekma. In this neighbourhood are meadows affording plenty of grafs, the rivers are abundantly ftored with fifh, and the woods replete with wild beafts. He might, indeed, have found places equally eligible before he came fo far to the north; but thefe were the refort of the Tungoofe, and he would have expofed himfelf to their depredations;

I

tions;

tions ; for the Afiatic tribes, as well as thofe of America, were inveterate enemies to each other, and fkirmifhes were the fure confequences of meeting in their hunting parties : even now thefe frequently happen.   While Omogai was in this fituation, two of his hunters fell in with a man of their own race, who was called Aley, or Eley, and had made his efcape from the Burati.   They took him to Omogai's who employed him as his labourer.   His remarkable ftrength, fkill, and activity, foon recommended him to Omogais' particular notice, and he was entrufted with the ma- nagement of fome excurfions.   The aftonifhing fuccefs that at- tended all his enterprifes, induced Omogai to make him overfeer of all his tribe and effects, which latter were confiderably increafed by the prudence of Aley's management ; and, in confequence of this increafe, the chief was obliged to extend his poffeffions to the vicinity of the prefent town of Yakutfk and the oppofite plains.

Omogai, who had a daughter by his wife then living, and a young woman whom he had adopted, was old, jealous, and dreaded the effects of Aley's power.   He obferved, that all his tribe efteemed Aley to adoration ; for they fuppofed him to poffefs fupernatural powers, and attributed his continual good fortune and fuccefs to the immediate influence of fpirits.   This made him uneafy ; and, with a view of fecuring his poffeffions and his name, he offered Aley his daughter in marriage.   Aley now avowed himfelf a Shaman, and affumed the powers of divi- nation.   He told Omogai, that his daughter would never have children, and therefore he would not take her ; but demanded the young woman that the chief had adopted, with whom he fhould have a numerous family.   The mother violently oppofed this union, but Omogai at length confented.   Aley's life was foon rendered very unpleafant by the perfecution of the mother and

daughter ;

daughter; but, having received very liberal prefents from Omogai at the time of obtaining the elderfhip of the tribe, and as rewards for his careful management not only of cattle and horfes, but alfo of men and women labourers, he was in poffeffion of independent wealth; and therefore retired, with Omogai's confent, two days' journey from his habitation, and eftablifhed himfelf on the plains 18 verfts north-weft of the prefent town of Yakutfk, by the fide of a branch of the river Lena, now dry. Here he remained till the death of his benefactor, when the greateft part of the tribe came over to him. Aley, who was now become extremely powerful, is reported to have had 12 fons and feveral daughters (Ghanghalas, or Chanhallas), the eldeft of whom was the founder of the Ghanghalafki tribe. The remaining hiftory of Omogai's wife and daughter I have not been able to learn; but the tribe is now known by the name of Batulinfk. It is faid to be about 300 years fince Omogai migrated hither.

The Batulinfki tribe was afterwards increafed by a number of the Chorintfi Burats; but the time of their union is unknown. I am inclined to think that their language was different; for if a Yakut be not immediately underftood by his brethren, he expreffes his diflike to repeat the fentence, by faying, " I fpoke not with the tongue of a Chorintfi." They know not whence they obtained the name of Yakut, but call themfelves Socha in the plural, and Sochalar in the fingular; I attribute the name to the founder of the town of Yakutfk, or to the name of the difcoverer of thefe people; for Yakutoff is no uncommon name among the Coffacs in the government of Irkutfk.

The firft intelligence that Ruffia obtained of thefe people was in 1620, when they were difcovered by the Coffacs that inhabited

bited the Mangazey. At that time they were divided into many tribes; and the diffensions that exifted among them contributed to their being fubdued.

Millach is the firft chief reported to have gone over to the Ruffians. He had a fmall tribe, which feparated from the Ghanghalafki, under the charge of the chief Tygin. Millach inhabited a hill on the eaftern fhores of the Lena, 60 verfts below Yakutfk, called Tfhebedal. He fupplied the Ruffians with food, and gave them 40 archers to fubdue Tygin and his tribe, which was accomplifhed on Tygin's falling in the field. Tribute was collected in 1630, and in 1632 the firft oftrog was built among them on the mountain Tfhebedal, but afterwards removed to the place where the town now ftands. Millach's tribe is now known by the name of Namfki Ulus, or Our Tribe, a name given to it by the Ruffians.

Their number is computed at Yakutfk to be 50,000 males; but I am inclined to think that they are not fo numerous; and my reafons are thefe: They fay themfelves, that in 1780 they were more numerous than they are now, much better circumftanced, and in an increafing ftate. At that time they had only one Commander and his Affiftant *. Upon the Socha chiefs bringing their annual tribute, they always obferved the cuftom of fhewing their particular attachment, by making thefe gentlemen a

* I am here induced to remark, that before the Emprefs Catherine II. eftablifhed governments and courts of juftice throughout the empire (1782), all thefe diftant towns and diftricts were governed by a voyavod and his fecretary, and Coffacs were fent among the tributary tribes to enforce the imperial mandates. On the eftablifhment of the government, every town had its mayor and different courts of juftice allowed; fo that there now exift many towns in thefe remote parts, where the inhabitants confift of government people only.

trifling

trifling prefent of furs, horfes, and cattle, and fupplying their table with flefh, fifh, milk, and butter, and alfo with wild fowl. Confidering their immenfe poffeffions, and the cheapnefs of all the articles, thefe prefents were never felt as of any confequence by the individuals who made them. At prefent, however, their ftock is confiderably diminifhed, not amounting to one tenth part of what it was. Inftead of having only the voyavod and his fecretary to deal with, they now know not how many commanders they have to pay their refpects to. A commandant, a captain of the diftrict, a director of economy, judges of the different courts of juftice, with their fecretaries and dependants, and other officers, are occafional travellers among them ; befide which, they complain of numberlefs exactions according to the arbitrary will of their fuperiors, only authorifed by their own prefumption.

Thefe circumftances undoubtedly difcourage the activity of the Yakut, who no longer endeavours to procure wealth, becaufe it is the likelieft means of making him the object of perfecution. Thus property, tranquillity, and population decreafe. The princes or chiefs dwelling near towns acquire their luxuries, and opprefs their dependant tribes to procure wine and brandy in addition to their koumis: this was never known among them till the year 1785. I will farther add, that in 1784 the diftrict of Giganfk produced 4834 tributary natives; but in 1789 their number amounted only to 1938. Mr. Bonnar, the captain of the diftrict of Zafhiverfk, told me, that the tributary nations in his circle amounted to only half the number that they were five years ago, and that thefe were very poor indeed. To my certain knowledge, upwards of 1500 Yakuti are hired as labourers by the inhabitants of the town of Yakutfk ; their wives dwell with the tribes, and do not fee their hufbands for years. However, I have fome other

Q   reafons,

reafons, which will be mentioned hereafter, to account for the decreafe in the population of this nation.

Of the firft huts about the river Newya, near Olekma, and all along the river Lena to its eftuary, the inhabitants are in indigent circumftances; as are alfo thofe of the Ochot, Amicon, Momo, Indigirka, Alafey, Kovima, and Jana, who felect fuch plains as afford food for their cattle; while the mountains are the places of refort of the Tungoofe. The Yakuti that inhabit the Vilui, Aldan, Ud, and all the intervening plains, are immenfely rich in cattle.

There is perhaps no nation in the world that can exhibit a greater variety with regard to fize. The affluent, whofe dwellings are fituated about the meadows on the fouth fide of the Virchoyanfki chain, are from five feet ten inches to fix feet four inches high, well proportioned, extremely ftrong, and very active; while the indigent inhabitants of the more northern parts are in general below the middle fize, indolent, and of an unhealthy complexion, evidently ftunted by the badnefs of their food, the feverity of the climate, and the want of proper cloathing. Their wealth confifts in horfes and horned cattle. The private property of no individual at this inftant exceeds 2000, all fpecies included; formerly, numbers of them poffeffed 20,000, according to their own teftimony, and that of the old Coffac before mentioned.

With regard to their capacity of fupporting themfelves, they are independent. Their only neceffaries are, a knife, hatchet (or palma), flint and fteel, and a kettle; and with thefe articles the all-providing hand of God fufficiently fupplies them, and capacitates them to furnifh the other tribes. From the iron ore of the

Vilui

Vilui they make their own knives, hatchets, &c. and of such temperature as baffles the more enlightened art of the Ruffians. This ore may be called native iron, from the little trouble they have in preparing it. Every utenfil and article of drefs they make themfelves.

In their roving parties, on the chafe or travelling, they only take with them a fcanty fupply of koumis, depending on chance for the reft; and fhould their purfuits prove unfortunate, they find their food in the inner bark of the pines and birch-trees, or the different edible roots. Squirrels are very good eating, but their favourite food is the whiftling marmot.

RELIGION.

The Socha regard themfelves as in a perfect ftate of demono-cracy. In general converfation, they call God, Tanghra; a church, Tanghra Dfhi, God's houfe; and Sundays, Tanghra Kuin, God's day. I could not obtain any explanation of the attributes of Tanghra. Thofe of other gods they explain as follows:

Aar Toyon (the merciful chief): To him they attribute the creation, and fuppofe him to have a wife, whom they call Kubey Chatoon (fhining in glory): they are both all-mighty.—Another god, named Wechfyt (the advocate), carries up their prayers, and executes the refolutions of the godhead: Wechfyt, they fay, ufed frequently to appear among them, and ftill continues now and then to fhew himfelf, affuming the form of a white ftallion, or different birds, from the eagle to the cuckoo.—Sheffugai Toyon (the protector): he intercedes for them, and procures all defirable things, as children, cattle, riches, as well as all good and com-

Q 2   fortable

fortable things: his wife they call Akſyt (the giver).—Theſe are their benevolent gods; and I may add to the number a being which they adore in the ſun: to theſe they offer ſacrifices only once a year. They attribute a particular being to the fire, and conſtantly offer ſacrifices, ſuppoſing him equally poſſeſſed of the powers of good and evil.—Their malevolent ſpirits are very numerous; for they have no leſs than 27 tribes or companies of aërial ſpirits: their chief they call Ooloo Toyon: he has a wife and many children: Sugai Toyon (the god of thunder) is his miniſter of immediate vengeance (Sugai is hatchet): the reſt they diſtinguiſh by the names of different colours. Cattle and horſes are ſacred to the different ſpirits whoſe colours they bear. They alſo reckon eight tribes of ſpirits inhabiting Mung Taar (everlaſting miſery). Their chief is called Aſharay Bioho (the mighty): theſe have wives, and the cattle ſacred to them are quite black: their departed ſhamans are ſuppoſed to unite to theſe. They are in great dread of another evil goddeſs, whom they call Enachſys (cowherdeſs): ſhe damages the cows, inflicts diſorders on them, deſtroys calves, &c. and is frequently honoured with offerings to be propitious to their ſtock.

CEREMONIES.

Their holidays commence with the month of June, and laſt about 15 days. The mares having caſt, a ſhort time is allowed the colts to ſuck, that they may acquire ſtrength; they are then tied up, or pent in coops about the hut, to prevent their ſucking at will; which is only allowed twice a day, when the mares are milked. The milk is collected in ſymirs, or large leather buckets formed like a bottle, wide at bottom, and narrow at the top, each

containing

containing about an anker; into this a fmall piece of the ftomach
of a calf or colt is thrown, and fome water mixed with it.   It is
then kept in conftant agitation by a broad-ended ftick, until it
ferments, and acquires an agreeable acidity, which is very nourifhing; and if taken in great quantities, it has an intoxicating quality.   Of this drink, which they call koumis, every one collects
as much as he can; and fome of the chiefs obtain more than
500 ankers of it.   A day is then fixed upon by each chief to
confecrate his ftock, which is performed as follows:-

A fummer hut is built of thin poles of a conical form, covered
with the inner bark of birch, on fome extenfive meadow.   It is
ornamented infide and out with branches of the birch-tree, and a
hearth is made in the centre.   Relations and acquaintances are
invited to the banquet; but all guefts are welcome of every nation indifcriminately.   The magicians take the head feats; others
are feated according to the eftimation of their feniority *.

When the hut is full, the elder fhaman rifes, and commands
one of the Socha that he knows to be qualified (namely, that has
not feen a corpfe within the month, and that has never been
accufed of theft, or bearing falfe witnefs againft any body, which
defiles them for ever, and renders them unqualified for this facred
and folemn tafk) to take a large goblet, called a tfhoron, which is
ufed to drink out of on folemn occafions, and fill it with koumis
out of the firft fymir; then to place himfelf before the hearth,
with his face to the eaft, holding the tfhoron to his breaft about

---

* Years do not fecure the title of fenior, (Oghonior,) which is the greateft term of
refpect that the Socha know.   Magicians have it, and all fuch as are capable of advifing
the proper means to be adopted to fecure fuccefs to fuch public and private concerns as are
virtuous and good.

two minutes. He then pours koumis three times on the hot embers, as an offering to Aar Toyon. Turning a very little to the right, he pours three times to Kubey Chatoon; then to the south he offers in the same manner to each of the benevolent gods. With his face to the west, he pours three times to the 27 tribes of aërial spirits; and three times to the north to the eight tribes of the pit, and to the manes of their departed sorcerers. After a short pause, he concludes his libation by an offering to Enachfys the cowherdess. The sorcerer then turns the man with his face to the east, and commences a prayer aloud, thanking the god-head for all favours received, and soliciting a continuance of their bounty. On concluding his prayer, he takes off his cap, with which he fans himself three times, and cries out aloud, " Oorui !" (grant) which is repeated by all present. The elder shaman then, taking the tshoron, drinks a little, and hands it to his brethren of the same order; from whom it passes to the company as they sit, except such as are defiled. Women are not admitted into the hut; nor are they, or the disqualified, allowed any of the koumis out of the first symir, which they call sanctified, as possessing the power of purifying and strengthening in a divine sense.

They all now go out of the hut, and seat themselves on the strewed branches of birch, in half circles fronting the east. All the symirs are carried out, and placed between the branches of trees stuck in the earth, and they commence drinking; every crescent having their symirs, tshoron, and presiding shaman, who fills the goblet, and pushes it about with the course of the sun. The quantity that they drink is incredible. Tournaments now begin, wrestling, running, leaping, &c.; and if any one carry off the prize in all the achievements, he is esteemed as particularly favoured by the deities, and receives more respect and

credit

credit in his teftimony than falls to the lot of a common man. When the ceremony is finifhed, they mount their horfes, forming half circles, drink a parting draught, and, wheeling round with the fun's courfe, ride home. Women attend, and form parties among themfelves at fome diftance from the men, where they drink, dance, &c.

MAGICIANS, OR SHAMANS.

Men and women are both admitted to this order; but very few of the latter, as particular circumftances attending their birth or infancy can alone authorife their inauguration. Young men are inftructed by an old profeffor, who accompanies them by day and night to the moft folitary parts of the woods; fhews them the favourite fpots of the fpirits of the air, and of the pit; and teaches them to cite their appearance, and claim their influence. I have heard moft wonderful relations of their power, even from the Ruffians; but, notwithftanding I have feen their enchantments or incantations many times, I never could difcover any of their feats equal to that of a common conjurer in England. The following is an account of their performance:

When a fick Socha fends for a fhaman to appeafe the wrath of the evil fpirits that torment him, the forcerer takes a fwitch, ties a few hairs from the mane of a horfe to the end of it, walks and jumps about the fick perfon, waves his fwitch, and conjures the demons to appear and relate the caufe of their tormenting him, and how they are to be appeafed.

After fome time has paffed in this invocation, he ftarts, pretends to fee the fpirits, and, liftening to their admonition for fome
time,

time, turns to the patient, and tells him whence the fpirits came; that it was with a view of deftroying him, but that they might be induced to accept as a facrifice, inftead of him, a fat mare or a cow, mentioning the particular colour. This is immediately procured; for whoever has one anfwering the defcription readily gives it.

The offering being procured, the fhaman dreffes himfelf in full form, walks with his fwitch to the poffeffed, embraces him, and commands the demons to leave him; then, rifing in great agitation, he fuddenly fprings upon the offering, raving and fhouting as much as he poffibly can: the beaft now ftarting, and being reftlefs, is a proof of the pain that it endures from the demon.

The following morning the facrifice is led to the place appointed, which is always on a rifing ground at the entrance into a wood. Four poles are driven into the ground, on which they erect a ftage covered with twigs, whereon the offering is flain and fkinned. The flefh is dreffed and eaten on the fpot; the bones collected, wrapped up in the twigs that were on the fcaffold, put infide the fkin of the animal, and ftuck at the top of fome tree on the fpot: if the facrifice was to the aërial fpirits, the head is directed upwards; if to the fpirits of the pit, the head is looking downwards. The forcerer then, arrayed in his magic robes, takes his tambour, and begins his formal fpells; beating his tambour, raving, jumping, and ufing an unintelligible jargon in the moft extravagant manner; his long hair hanging over his face, he conjures the fpirit of the facrifice to its demons, and the demons to their proper place of retirement; feems, feveral times during his incantations, to faint, during which paroxyfms he receives the infpired power of prognofticating the fate of the difeafed, and the day either of his reftoration to health, or of his death. If

he

he prove miftaken, it is not regarded as arifing from want of fkill, but the unacceptablenefs of the facrifice, which is occafionally re-newed till he dies or recovers.

If a fhaman acquaints any family that fome demon is intent on inflicting a punifhment, offerings are made to avert the evil ; not of live beafts, but the rich fkins of animals, which are hung up in a confpicuous part of the hut, and buried with the owner when he dies.

The magician's drefs is a leather jacket, with fleeves from the fhoulder to the elbow ; along the outer feam, long flips of leather are fewn, as alfo round the bottom, hanging to the ground. The jacket is covered with iron plates, and pieces of iron and brafs hanging, which makes a difmal noife while he is leaping about and beating his tambour. He has alfo a piece of leather, like a long apron, reaching from his chin to his knees, tied before, and ornamented in the fame manner. His tambour is very large, and alfo ornamented in the edges and crofs bars with iron and brafs ; and his ftick is covered with the fkin of fome fhort-haired animal. He alfo wears, at the commencement of his incantations, a fur cap ; but this he throws off almoft as foon as he begins his magic fpells.

METHOD OF DIVIDING TIME.

The year they call gil, and divide it into four feafons, faas (fpring), foyin (fummer), kuifan (autumn), and kifun (winter). A month they call ooi ; of thefe they have 12 in the year, of 30 days, adding a fupplemental moon for the deficiency every fixth year. Their months are named as follow :

R   Befia

Befia	Budding month	May
Otti	Hay ditto	June
Otterfhachia	Hay forks ftacking	July
Tierdinnai	Fourth	Auguft
Beffinnai	Fifth	September
Altidnai	Sixth	October
Settinnai	Seventh	November
Okfinnai	Eighth	December
Tochfinnai	Ninth	January
Ollunnai	Tenth	February
Koluntutor	Stallion	March
Buffuftur	Thawing	April.

They have no fpecific name for the fupplemental month every fixth year.

They know the time of night by the fituation of the great bear and the polar ftar : the former they call araghas folus.

Of the approaching feafons they judge by the following phenomenon. If the pleiades, which they call oorgel, appear before the moon when feven days old in the month of January, they expect fpring to commence in the beginning of April; if when nine days old, at the end of April; but if this happens on the tenth day, they expect a late fpring, and begin to be very faving of their fodder. They reckon diftance by time; and 30 or 40 verfts, according to the goodnefs of the roads, make a day's journey.

PUNISHMENTS, CUSTOMS, SUPERSTITION, &c.

I have not traced any atrocious vices among the Yakuti or Socha. Robberies are feldom committed; fometimes, indeed,

they

they lose cattle; but this I believe to be more the effect of their straying than their being stolen, as detection is almost certain; for they have an astonishing memory, and relate all their losses at every public meeting, as weddings, &c.; and if any one has in his travels seen such a beast as is described, he relates where and when: thus they are traced, and the punishment is, not only restoration to the party injured, but the thief is compelled to make good all the losses of the other Yakuti during the year, whether he has stolen the property or not.   If an accusation be laid against any by his companion, of having stolen and eaten or killed cattle, he must either pay for the same, receive a flogging (which is very disgraceful), or take an oath of his innocence; and, should an innocent man be accused, he will, in general, rather pay for them than take the stipulated oath, which is administered with the following ceremony:

A magician places his tambour and dress before the fire, the embers of which are burning. The accused stands before it, facing the sun, and says: " May I lose during my life all that " man holds dear and desirable, father, mother, wives, children, " relations; all my possessions and cattle; the light of the sun, " and then my own life; and may my spirit sink to eternal mi- " sery (mung taar), if I be guilty of the charge laid against me!" The magician throws butter on the hot embers; the man accused must then step over the tambour and dress, advance to the fire, and swallow some of the exhaling smoke from the butter; then, looking to the sun, say, " If I have sworn false, deprive me " of thy light and heat." Some of the tribes close the ceremony by making the accused bite the head of a bear; because they allow this beast to have more than human wisdom, and suppose that some bear will kill the aggressor.

They are very revengeful of infults; nay, even entail revenge on their progeny: nor do they ever forget a benefit received; for they not only make reftitution, but recommend to their offspring the ties of friendfhip and gratitude to their benefactors. They are very obedient to their chiefs and oghoniors, and fhew their attachment by frequent vifits and prefents. They are extremely hofpitable and attentive to travellers, efpecially to fuch as behave with a degree of good nature, and very inquifitive and intelligent; for they afk queftions freely, and anfwer any without embarraffment or hefitation. They are anxious to fecure friendfhip and a good name, and feem to ftudy the difpofitions of fuch as may be of fervice to them, to whom they are liberal in prefents, and even in flattery. They deliberate in council on all matters of public concern, as the courfe to be taken by each in the chafe, &c. The oghoniors are furrounded by the reft, and their advice is always taken. I have never feen an old man contradicted or oppofed, but always as implicitly obeyed as a father of a family. A young man ever gives his opinion with the greateft refpect and caution; and even when afked, he fubmits his ideas to the judgment of the old.

The Yakuti are a healthy and hardy race, bear the extremes of heat and cold to an aftonifhing degree, and travel in the fevereft frofts on horfeback, frequently fuffering much from hunger; they are, however, fubject to rheumatic pains, boils, the itch, and fore eyes; and great numbers were carried off in 1758 and 1774 by the fmall-pox and meafles: the fhamans are their doctors. They are extremely fuperftitious, and almoft every tribe has its object of veneration, but not of worfhip, as the eagle, the fwan, the ftallion, &c. Ravens, crows, and cuckoos, are ominous birds; if thefe perch near their huts, they dread fome misfortune, which

is

is only to be averted by fhooting the bird. Eagles and large birds of prey are, on the contrary, the foreboders of good. They always take care that the doors of their huts fhall face the eaft. The fire-place is nearly in the middle; the back of the chimney towards the door, and a free paffage quite round it. The fides of the hut are furnifhed with benches and fmall cabins, which ferve for fleeping places. The bench extends about four feet into the hut, where the inhabitants fit. The men keep the fouth fide, and the women the north. Except the hoftefs, no woman may give any thing to eat or drink to a male ftranger before the fire-place, but muft walk round the chimney to prefent it.

They never wafh any of their eating or drinking utenfils; but, as foon as a difh is emptied, they clean it with the fore and middle finger; for they think it a great fin to wafh away any part of their food, and apprehend that the confequence will be a fcarcity. Their earthen veffels they keep extremely clean, becaufe they can make them fo by burning, in which cafe the fire accepts the remains that adhered to the fides. Before they begin to eat any thing, they throw a fmall fpoonful into the fire as an offering of thanks. The rich are efteemed to be under the protection of the gods; but the poor are rejected as forfaken, and only protected by their relations, or obliged to hire themfelves out to work. Every Yakut bears two names, and is never called by the right, except in cafes of neceffity; thus they think they evade the fearch of the evil fpirits bent on tormenting them. They never mention the dead, except allegorically, and leave the hut to ruins wherein any one has expired, thinking it the habitation of demons.

Travelling

Travelling with Captain Billings to the river Mayo, I obferved the following offering in the hut of Surtuyea Birdugin, an unchriftened prince or chief, to Sheffugai Toyon and his wife Akfyt, for the bleffing of children, called Ogo Oyetto, the child's neft. It was a horfe-hair cord tied round the chimney, leading to the fleeping place of the hoft and hoftefs, ornamented with bunches of horfe hair. Two round pieces of the bark of birch, to reprefent fun and moon, fufpended; alfo the reprefentation of a ftallion and a mare formed out of the bark, and a few wooden dolls dreffed. The cord was faftened to the poft at the head of their bed, where was placed a wicker bafket with mofs and fur at the bottom; and on a little table in the bafket was fet a very fmall wooden bowl, containing boiled flour; all of which was placed by a fhaman with great magic ceremony. The prince has three wives, and was married 15 years without having had any children before this offering was made; but afterwards each of his wives bore their fhare of children, and he has now fix fons and daughters. This account was related to Captain Billings and me in the prefence of Mr. Hornoffky, the captain of the diftrict of Yakutfk, by the prince himfelf: each wife has her feparate dwelling fome miles diftant from either of the others; and a fimilar offering is placed in each of their huts.

At the time of parturition, the hufband is called, and two fkilful women in his prefence affift the delivery. If a fon be born, a fat mare is killed on the third day; all the neighbours are invited to fupper; the child is rubbed all over with fat, and a name given to it,—the more infignificant the better, for an elegant name would entice the demons to be continually about it. No ceremony is obferved if the child be a daughter.

MARRIAGE

MARRIAGE CEREMONY.

The ceremony of buying a wife is extremely formal and tedious. A young man who wifhes to marry fends his friend to afk the confent of the bride's father, and what kalym (purchafe) he demands; that is, how many horfes and cattle, as alfo the quantity of raw meat, horfe flefh, and beef, that he requires for treats and feafts : this they call kurim; half of the quantity is always given in prefents to the bridegroom by the bride's father, and is called yrdy. The daughter's inclinations are always confulted; and, if fhe does not object, the kalym and kurim are ftipulated. The bridegroom kills two fat mares, dreffes the heads whole, and the flefh in pieces, and goes with three or four friends to the father of the bride. On his arrival at the hut, one of his friends enters and places one of the dreffed horfes' heads before the fire, and returns to his companions without fpeaking a word. They then all enter the hut, and, a forcerer being placed oppofite the fire, the bridegroom kneels on one knee with his face towards it, into which butter is thrown; he then lifts up his cap a little, and nods his head three times without bowing his body. The forcerer pronounces him the happy man, and prophefies a fucceffion of happy years, &c. Then the bridegroom rifes, bows to the father and mother, and takes his feat oppofite the bride's place, but keeps filent. The meat is then brought in, and the father of the bride diftributes it among his own friends, but kills a fat mare to treat his new guefts. Supper being over, the bridegroom goes to bed; the bride, who has not been prefent, is conducted into the hut and to his bed by fome old woman, and they fleep together; fometimes, however, the bride does not appear at the firft vifit. In the morning, the friends return home;

but

but the bridegroom remains three or four days. A time is now fixed for the payment of the kalym, either at the new or full moon. The kalym and kurim are then carried, without any ceremony, and delivered in the prefence of many friends, who are feafted, and the bridegroom remains again three or four days, and fixes a time to receive the bride at his own dwelling, which muft be new built on purpofe, and this alfo at the new or full moon. All her relations, male and female, with friends and neighbours, fometimes more than a hundred, accompany the bride with her father and mother, taking with them eight or ten fymirs full of melted butter, and the dreffed meat of three fat mares. They go to the new hut prepared for them ; three men are fent to the bridegroom in his old hut, and the greateft drinkers are chofen for this purpofe. On entering, the firft fays, " We are " come to fee your dwelling, and to fix pofts before your door." They then kneel on one knee before the fire. An ayach * is filled with koumis, and handed by two men to the thrèe kneeling, each of whom empties an ayach at three draughts. They then rife and go out, all the company faluting them with one cheer. Three others enter ; the firft with nine fables, the fecond with nine foxes, and the third with 27 ermine fkins : thefe they hang on a peg in the chief corner of the hut, and retire. Then a number of women conduct the bride, her face being covered with ermine fkins, to the hut ; the entrance has a wooden bar placed acrofs it, but of no ftrength, which the bride breaks with her breaft, and enters the hut. She is placed before the fire, holding her hands open before her, into which feven pieces of fticks are put ; as alfo feveral pieces of butter, which fhe throws into the fire. The fhaman pronounces a bleffing ; fhe then rifes, and is again con-

---

* An urn-fhaped wooden veffel with three legs, which contains from two to four gallons.

ducted,

ducted, with her face concealed all the while, to the new hut, where the cover is taken from her face. The bridegroom enters, and feasts his guests two days; then presents all his relations with cattle, over and above the kalim; which is, however, returned, on paying their formal visits, perhaps a year or more afterwards. Polygamy is allowed, and some have six wives; but the first is respected by all the rest; they dwell in separate huts; and in case of bad conduct they are returned home, and the greatest part of the kalim is given back. This, however, very seldom happens: I have not been able to hear of a single instance.

BURIALS.

The corpse is first dressed in the best apparel of the deceased, and stretched out; the arms tied tight round the waist; then inclosed in a strong box, with the knife, flint, steel, and tinder; also some meat and butter, " that the dead may not hunger on the road to the dwelling of souls." A shaman presides; the wives and relations accompany the procession to a certain distance; the favourite riding-horse of the deceased is saddled and accoutred, with hatchet, palma, kettle, &c, and led to the place of interment, as is also a fat mare. Two holes are dug under some tree; then the horse is killed, and buried in one, while the corpse is laid in the other. The mare is killed, dressed, and eaten by the guests; the skin suspended on the tree, under which the body lies with the head to the west. The shaman takes his tambour, and invokes the demons to let the spirits of the departed rest in peace, and finishes the ceremony by filling up the grave. A shaman is buried with the same ceremony, and his tambour with him. If an elder brother die, his wives become the property of the

S   younger;

younger; but if a younger brother die, his wives are free; yet they seldom marry again, except they be very poor.

Their dress is much more complete than that of the Tungoose; and the more wealthy among them wear a cloth coat lined and trimmed with fur, with tight and well made pantaloons; but their boots are ill shapen. The women dress very like the Tungoose, but are in general not so clean or sprightly.

EMPLOYMENT.

About the 25th of June, at the conclusion of their holidays, they commence their summer occupation by collecting a great supply of the inner bark of the pine and birch, which they dry on racks in their huts: hay-making and fishing then occupy the time till berries are ripe, when they collect an immense quantity, and boil to preserve them. In the beginning of October they kill their winter stock of cattle for food, and let it freeze, which preserves it fresh and good; of course they save so much hay. In October and November they catch fish under the ice. Toward the end of the latter month, they go out on the chase: for wolves and foxes they place in their tracks poisoned baits of corrosive sublimate, which they call sullima, and also of nux vomica (Tshillebucha); besides having traps set, spring bows, &c. They are very expert archers, and have a plentiful supply of arrows in their quivers.

The women make all their cloathing, look after the cattle, milk the cows and mares, chop wood, dress food, &c.

They

They have no amusements beyond feasting, eating, and drinking. Sometimes, indeed, the women dance, which, however, is only forming a ring, and walking round with the sun's course. Their songs are inharmonious, and almost all extempore on any object that strikes the imagination.

They dress leather for use in the following manner: For symirs, they take a fresh skinned cow's or horse's hide, and steep it in water a few days, when the hair easily rubs off. It is then hung up till nearly dry, when they lay it in blood until soaked through, and then hang it in a smoky place for a considerable time: of this they make their buckets and soals of boots, &c. as the latter are completely water-proof, and the buckets, or symirs, even retain oil. The legs of boots they make of colt's or calf's skins, scraped and rubbed till they be soft, then sewn, steeped in blood, and dried in smoke; afterwards blackened with wood-coals and fat several times, and smoked again; they then are water-proof. Elk and deer skins are dressed with and without the hair on, by being covered with a paste made of clay, and the undigested food from the maw; or with cow-dung hung up till nearly dry, then rubbed and scraped till soft. They are then either kept of their natural colour, or dyed of a red colour with the bark of alder and ashes boiled together, or else of a yellow colour with the roots of sorrel. The thread with which they sew their clothes is made of the sinews from the legs of the horse, deer, or elk.

Notwithstanding the strictest enquiries, I could not obtain any intelligence of remarkable places, or springs, of any kind, except the mountain where Commodore Bering obtained coals in his expedition of 1725, and forged his anchors near the famous mount

Tshebedal,

Tſhebedal, from whence they were tranſported by water to Yudom-ſky Kreſt, carried by land to Urak Plotbiſha, and down the Oorak, or Urak river, to the ſea and port of Ochotſk.

The mountain is ſituated on the Yakutſk plains 60 verſts north of the town, on the confines of the Lena; it is called by the Ruſ-ſians Surgutſkoi Kamen. I paid a viſit to it in the beginning of March 1789; and found it the extremity of the ridge that bounds the plains toward the river; it is perpendicular; about 50 fathom high; formed chiefly of iron-ſtone, free-ſtone, and ſtrata of coals, lying horizontally, from one to about three and a half feet thick, and of inconſiderable length; they then break off, and the ſame ſtrata ſeems continued ſometimes five or ſix feet higher or lower. They reſemble petrified trees, the end towards the north being thicker than towards the ſouth; ſome have ſhort branches ſhooting from them of about five or ſix feet in length. About half way up the mountain, there ſeems in one place to be a warm ſpring; for I obſerved a vapour or faint ſmoke aſcend from it, and the ground near it was wet.

On my way thither, I paſſed the night in one of the huts of a Yakut about 10 verſts from the mountain, and obſerved there a ſmall furnace with a pair of hand-bellows fixed, which were double, and gave a conſtant blaſt when worked with both hands. I ſaw ſome ſpecimens of iron ore obtained in the neighbourhood, exact-ly reſembling that of the Vilui in curious forms and ſhapes. This iron my hoſt worked into knives, palmas, hatchets, &c. without fuſion, bringing it into a very ſoft ſtate by heat, and beating it out. He uſed charcoal for this purpoſe, nor did he know, till I ſhowed him, that the coals, which he called black ſtones, would burn; and he was inclined to think me a ſorcerer for making
                                                                them

TO THE EASTERN OCEAN, &c.

them inflammable. This mountain, however, afforded him grind-ftones.

Returning homeward, I arrived late at a hut about 18 verfts from town, and refolved on paffing the night in it. The land-lord, an old Sochalar, entertained me with an account of his own pedigree; tracing himfelf, in a direct line, from Aley; and af-fured me that this was the neighbourhood to which he retreated from Omogai; of which retreat he gave me the following account:

Aley received numberlefs prefents from Omogai and his de-pendents during his elderfhip; but was obliged to leave all thefe behind him, and was driven from Omogai with only two old mares, on which he and his wife rode; all his poffeffions at the time were, the clothes on their backs, a bow and arrows, a hatchet, palma, and two knives, with fire materials. Aley thought this a convenient fpot; and, halting the fecond day, built a temporary hut, collected carefully the dung of his mares, and, when the wind blew towards Omogai's habitations, made fires of the dung, the fmell of which allured the ftrayed cattle to his dwelling: he then carefully fed and watered them, and drove them back.

Aley now built himfelf a very large hut and ftorehoufes. Be-fore his hut he ftuck up pofts, with carved tops, for travellers to tie their horfes to, and made a number of hurdle coops and pens clofe to his dwelling. The cattle conftantly returning, with frefh numbers of milch cows and mares, Aley collected immenfe quan-tities of butter, milk, and koumis; and, having been very fuccefs-ful in the chafe, he had a great fupply of the meat of the elk and deer, with game of all kinds. He now purpofely wandered to
the

the places that he knew to be frequented by Omogai's tribe, fell in with fome of his hunters, and brought them to his dwelling, having previoufly cautioned his wife to keep the ftrayed cattle far from his habitation. Omogai's people were aftonifhed at the elegance of the habitation, and the profufion of fifh and flefh of different animals; but, above all, at the quantities of koumis and butter with which he treated them, knowing that he had no cattle.

Aley told them, that he had been admonifhed by his fpirits, or demons, to form the different pens which they faw before and round his hut, and to affix the pofts for the horfes of his vifitors; affuring him, that his guefts fhould be numerous, and his pof-feffions great. He punctually obeyed the injunction of his de-mons; and, to his aftonifhment, obferved a white-mouthed ftal-lion lead to his pens a number of mares and cows: thefe his wife milked; which being effected, they vanifhed from his fight, but returned every night and morning. He kept his guefts all night; and in the morning fent them away, with provifions for the road, and prefents of rich furs for Omogai, his wife, and daughter.

Aley had now feveral children, and was very defirous of ob-taining, by fair means if poffible, the property that he ought to have received from Omogai. Not doubting that the reports of the hunters, and the prefents fent, would effect a reconciliation between them, and aftonifh his whole tribe, Aley refolved on pay-ing his old mafter a vifit, with additional prefents, and inviting him, with the heads of his tribe and families, to pafs a day or two at his habitation. He was well received, and Omogai pro-mifed to attend, with his wife, daughter, and friends, at an appointed

pointed time. Aley, on his return, built a very large temporary hut to receive his guefts. They came and brought with them a confiderable prefent of cattle. Aley treated them with great fplendor for three days, received the prefent as fuch, and claimed with humility the cattle and labourers that had been unjuftly kept from him, and which were his due, becaufe he had ferved for them. Omogai acknowledged the demand to be juft; but by the oppofition of his wife and daughter was prevented from granting it. They departed with their friends; but Aley, through the influence of his demons, caufed a violent ftorm and extreme darknefs, in which they ftrayed feveral days; at length Omogai, his wife, and daughter, reached home; but the greater part of his friends returned to Aley, and acknowledged him their chief; being much difgufted at the refufal that he had received, and perhaps dreading the influence of his fupernatural powers.

Omogai died fhortly after, and the greateft part of his tribe went over with their cattle to Aley; but Batulin, one of his ftewards, married the daughter, and fecured the reft. She never bore him children, however; but Batulin took other wives, and had feveral.

Aley had twelve fons and feveral daughters. Changhalas he initiated in his magic art, and he was the founder of the Changalafki tribe.

This account is general among the Changalafki, who adore the ftallion; but the hiftory, as I have before related it, is credited by the greater number.

To

To give my readers an idea of the population of these northern parts of Siberia *, I here note the inhabitants from the latitude of 64 to the extremity of the north coast, and from the river Kovima, westward to the Anabara.

The district of Zashiversk comprehends the rivers Kovima, Alasey, Indigerka, and Yana, and those that flow into them ; the tributary nations are, Yakuti - - 2810
Lamut and Tungoose - 742
Yukagiri - - 322
Tshuvantsi and Chatinsy - 37

Tribute received 1788 amounts to rubles 4560 for 3911 males.

The circuit is about 6000 versts in circumference. The district of Gigansk, a town north of Yakutsk on the Lena, contains one church, two government houses, seven private ones, and 15 huts. It has a mayor (Gorodnitshik) and his chancery, a court of the district (Zemikoi Sud), and a magistracy, although the merchants are mere trading pedlars, and only two, I think, in number. Its circuit also is about 6000 versts from the Yana to the Anabara, which divides the governments of Tobolsk and Irkutsk. The tributary nations are, Yakuti - - 1449
Tungoose - - 489
                                                       1938

Tribute received in 1788—56 sables, 262 foxes, and rubles 1169 in money.

The Russians inhabiting both districts, including exiles, &c. do not exceed 750 males.

* All Asiatic Russia, east of the Uralian or Virchoturian chain, is now called Siberia.

CHAP. XI.

*Leave Yakutſk.—Arrive at the Village Amginſkoi.—Uſt Mayo Priſ-
tan.—Arrive at Ochotſk.—Two new Veſſels launched, and named
The Glory of Ruſſia, and The Good Intent.—The latter Ship
wrecked: a Circumſtance which had been predicted by the ſuperſti-
tious Inhabitants from an ominous flight of Crows.—A Courier
from Peterſburg arrives.—An Iſland diſcovered, and named
Jonas's Iſland.—Arrive at the Harbour of St. Peter and St.
Paul in Kamtſhatka, where we winter, making only occaſional
Excurſions.*

THE ice of the river Lena broke up the 17th May, and on the
22d we croſſed to the Yarmank, where horſes were provided for
us, and we were attended by the Iſpravink of Yakutſk. The
river had overflowed the low country; ſome ice was ſtill floating down the ſtream, and a great number of trees.

We immediately proceeded on our journey to the Aldan, at
the diſcharge of the river Mayo. I have already deſcribed the
plains between Yakutſk and this river; but on our preſent route
we ſtopped at a village called Amginſkoi Sloboda, inhabited by
168 Siberian coloniſts, ſent hither to grow corn, which, however, does not anſwer, except for their own ſupport, and not always that; for in ſome years nothing is produced. The inhabitants get their bread chiefly by trading with the neighbouring
tribes in trinkets and brandy. They informed us, that none of
the wandering Tungooſe were yet arrived at the Uſt Mayo; and,

T   upon

upon being confulted about the road to the Aldama and Ulkan rivers (at the difcharge of which Captain Billings had promifed to meet Captain Saretfheff), they gave fo bad an account of it, that Captain Billings refolved to give up the thoughts of taking this road. He therefore difpatched a Coffac with an order from the Ifpravink to the Yakuti inhabiting the plains, that they fhould fend immediately to the Aldan Stanok 16 horfes for our conveyance by the old road to Ochotfk.

On Thurfday the 31ft of May, we arrived at the Uft Mayo Priftan, oppofite the difcharge of the Mayo, and immediately fent a man to the prince of the Tungoofe, who refides about ten verfts up the Aldan. This is the chief or head of all the Tungoofe, who has a number of Yakuti under his direction. He has feveral wives of the Yakut and Tungoofe, is by both thofe nations much refpected, and acts as an agent to the Mongal Tartars on the Chinefe frontiers, to the Yakuti, and the Tungoofe. He came to us early in the morning of the 1ft June, and told us, that the road which Captain Billings purpofed travelling would be attended with fome difficulty; that the deputies, or elders, of the wandering tribes were not yet arrived; that he would fend a letter to Captain Saretfheff, and anfwer for its being delivered in 20 days, if he came near the coaft about the eftuaries of the Ulkan or Aldama rivers. In confequence of this, Captain Billings difpatched a letter to Mr. Saretfheff, defiring that he would return immediately to Ochotfk, and meet him there, as he hoped the fhips would be ready for fea.

Boats were procured, and on the 4th June we fell down the ftream of the Aldan, 150 verfts to the Old Aldan ftage, where we

arrived

arrived on the 7th, at fix o'clock in the evening, having for the laft eight days had rainy and ftormy weather.

The ordered horfes were not yet arrived, nor was the Coffac who was fent for them; but we obtained twelve ftage horfes, with which we proceeded to Ochotfk on the 8th at noon, and arrived at the port on the 21ft. Here we found the largeft fhip ready for launching, and the other nearly fo. All the articles arrived fafe, and all hands in good health and fpirits; and toward the end of the month Mr. Saretfheff returned, having received the letter fent him from Uft Mayo Priftan.

Dr. Merck, our naturalift, was making a collection of the curiofities about the Mariakan mountains; but orders were difpatched for his return, as we expected to get to fea about the middle of Auguft.

Toward the middle of July, our largeft fhip was launched; fhe went off the ftocks extremely well; but, owing to the fhoals in the river, it was almoft three weeks before we could get her into deep water near the difcharge of the bay, where fhe took in a part of her cargo. She was then taken out to fea about five miles, over the fand banks, and brought to anchor in fix fathom water, with a bottom of fand and ftones. We employed the tranfport galliots to carry guns, ftores, &c. on board while in this fituation; for fhe could not have paffed the fhallows even in proper ballaft. She was named, by order of the Emprefs, the Slava Roffie, Glory of Ruffia.

On the 8th Auguft, the fecond fhip was launched, and called the Dobroia Namerenia, Good Intent. She was rigged, and ready

T 2

to go out early in September; it was, however, neceſſary to wait for the ſpring tides to carry her over the grounds; a galliot was loaden with her ſtores and ammunition, and got ready to accompany her out.

In the evening of the 7th of September, Captain Billings reſolved on carrying the ſhip out the next morning; Mr. Loftſoff, the pilot of the port, was ordered to take the charge, and get all the boats belonging to the port manned, and in readineſs, that, in caſe the wind ſhould fail, they might tow her out; the boats of both ſhips were alſo ordered to attend. Captain Hall, who had the command of this ſhip, ſlept on board. At ſix o'clock in the morning of the 8th, I went on board to get a book out of the cabin. Before I got up the ſide of the ſhip, the Captain aſked me whether I brought any orders to go out. I told him that I did not, and aſked him if he thought it was poſſible. The wind was favourable, but very ſcant; a heavy ſwell from the ſouthweſt right on ſhore, and the ſea breaking amazingly over the banks, and on the beach: this I thought indicated a ſouth-weſt breeze, beſide which, it was very foggy. Captain Hall ſaid, he thought it impoſſible, and certainly ſhould not go out, unleſs the commander came on board himſelf, and inſiſted upon it. Mr. Koch, the acting commandant, who was on board the tranſport veſſel cloſe aſtern, aſked Captain Hall, if he ſhould follow him? The anſwer was, " No, unleſs you mean to be caſt aſhore; but I ſhall not go myſelf if I can help it." At half paſt ſeven Captain Billings came on board, and, after ſome converſation with Mr. Hall, ſaid, " The pilot ſhall determine." The pilot arrived; Mr. Hall repreſented his fears; and added, that Mr. Loftſoff was, perhaps, not aware of the difference between the ſhip that he was then taking charge of, and a galliot of ſixty tons. Captain Billings

lings faid, he did not think the danger fo great as Captain Hall reprefented; and urged the neceffity of getting out this fpring-tide if poffible, owing to the late feafon, and his defire of paffing the winter on the north-weft coaft of America. The pilot affirmed that there was not any danger. Captain Hall then told the crew to obey every order of Mr. Loftfoff, and refufed to have any charge in carrying the fhip out, but proffered every affiftance in his power. At eight o'clock, high water, juft as the tide was turning, tow-lines being given to fix barges and boats, fhe caft off; it was a perfect calm; the fwell very heavy; and the fea breaking over the banks with great violence. The boats towed her through the paffage, keeping her head againft the fwell; but when the firft boat (the largeft, which had, I believe, fixteen oars), got into the breakers, fhe fhipped a heavy fea, and caft off her tow-line *. The fhip pitched exceedingly, and the fkiff along fide had her head carried under water, being entangled with the fore-chains, and two men were wafhed out of her. Every boat now caft off her tow-line to fave the men, one of whom only they picked up. The fhip, driven at the mercy of the fwell, ftruck on the beach, and ftuck faft. At a quarter paft nine, a light breeze fprung up from the fouth-weft. Her mafts were cut away, but to no purpofe; for the tide was on the ebb, and fhe was foon left dry. There was no time to be loft; the wreck was cleared away, and as much of the rigging and ftores carried on fhore as poffible. The refo-lution was immediately taken to fail in one fhip to Kamtfhatka, and there build a fmall veffel, during the winter, of the materials of the loft fhip. Not having time to break her up, it was alfo refolved to burn her, as the quickeft method of getting at her iron-work. This was put in execution on the 9th September.

---

* This boat rowed on board the other fhip over the bar at anchor, and returned with the flood tide: fhe could not turn in the breakers.

I

The wind was this morning moderate from the fouth-weft; but the furf beat with fuch violence againft the beach, that the fpray nearly reached the church. It frefhened in the afternoon, and died away about midnight.

The lofs of this fhip had been foretold by the *fuperftitious* inhabitants of the town, from the following remarkable circumftance: In the fpring of the year, a flight of crows were fighting in the air, and making a dreadful noife. One of them was killed by the reft, and fell upon the deck of this fhip. The whole fwarm immediately defcended, and entirely devoured the vanquifhed bird, leaving no other veftiges than the feathers behind. This very remarkable occurrence, which was related by all our officers, workmen, and inhabitants, happened while I was at Yakutfk.

I have judged it neceffary to be particular in my account of this very unfortunate circumftance; and fhall only add, that it appeared to me very fortunate that the fhip did not get into the breakers in the narrow channel; in which cafe fhe muft inevitably have been driven on the bank, and in all probability not a foul on board would have been faved. Captain Saretfheff was on board the Slava Roffie, at anchor five miles out at fea, without a boat.

The morning of the 10th would have been very favourable, with a leading wind from the north till 11 o'clock, when it veered to the fouth-weft. I was fent on board the Slava Roffie with ftores, an anchor and cable, in the long boat. Captain Saretfheff lamented his not having been on fhore, to have oppofed

the

the carrying out the fhip under fuch inaufpicious profpects. He alfo expreffed great regret at the fhip's having been burnt; but was happy to hear that no lives had been loft, except that of one man, who had neither wife, nor family, nor friends. The body of this man had been feen floating at no great diftance from the fhip, which filled every one on board with melancholy fentiments. I returned in the evening with the tide. The next day Captain-Lieutenant Bering went on board, and in the evening Captain-Lieutenant Saretfheff returned.

September 14th. A courier arrived from St. Peterfburg, bringing intelligence of the war with Sweden, recommending the greateft economy in our proceedings, (as money was extremely fcarce in Ruffia) and ordering the expedition to return to St. Peterfburg if we had not failed from the port of Ochotfk, or if things were not in complete readinefs for fea; for they experienced a great want of naval officers and men. Towards evening it blew frefh from the fouth-weft, and precluded all communication between the parties on board and on fhore.

On the 15th, it blew a hard gale from the fouth-weft. We feveral times obferved from fhore, that the fhip drove, and towards night that her top-mafts were ftruck. The gale continuing, we made fires along the beach, and obferved a lanthorn on one of the maft-heads.

The 16th, the gale continued with unabated violence, and we faw with our glaffes that the fhip had three anchors a-head; notwithftanding which, fhe frequently drove, and we expected her every moment on fhore. She had very few hands on board and

no

no boat. The night was terrific, with very heavy rain. We again made fires all along the beach.

On the 17th our anxiety increafed with the gale. We did not obferve her drive; but frequently the thick weather hid her from our fight: towards evening, however, to our great joy, the gale abated. Very early in the morning of the 18th, we fent on board all neceffary hands, ftores, materials, &c. employing all the boats. We obferved that the fhip had driven above a mile to the north-eaft into three and a half fathom water; and had fhe gone about forty fathom further, fhe would have got on a fhoal.

On the morning of the 19th we had four inches of fnow on the ground. Captain Billings and all abfolutely neceffary hands embarked *; and about noon we weighed anchor, with a moderate fouth-weft breeze, fhaping our courfe to the eaft of the fouth. On the 22d we faw an ifland bearing fouth-weft diftant about forty miles, with detached rocks round about it. We hove the lead, and found bottom with twelve fathom line only. This fmall ifland is not noted in any of the charts, and we gave it the name of Jonas's Ifland. On the 28th we paffed the remarkable mountain called Alaid, rifing out of the fea, and terminating in a cone. Some on board declared, that in clear weather they had feen it at 350 verfts diftance: its fituation, however, is about twenty miles from the fouth point of Kamtfhatka. The fame day we paffed the Kurillan ftraits between the fecond and third ifland, and arrived at St. Peter and St. Paul the 1ft day of October.

* We were compelled to leave feveral behind us, with ftores, &c. who were to follow us to Kamtfhatka, with the tranfport veffel, in the fpring.

Nothing

Nothing interefting, or worthy of notice, happened in this trip. We had very boifterous weather, and a difagreeable fhort fea until we came into the Northern Pacific Ocean, where we experienced an amazing difference in the climate, which was mild and pleafant. In Kamtfhatka the weather was very agreeable. The kitchen gardens belonging to the Coffacs were full of cabbages and other vegetables, and the views around were more beautiful than any thing of the kind that I ever remember to have feen. The looks of all the inhabitants feemed to evince health, plenty, and content: and, in fhort, every thing was completely the contrary of what we had feen and felt on the river Kovima.

We unloaded and unrigged our fhip, built barracks for our men, and ourfelves occupied the houfes of the inhabitants, three or four officers being ftowed in a fmall room; but before the winter fet in we made additions to the buildings, lived very comfortably without diftreffing the inhabitants, and in perfect harmony with them. Fifh and wild fowl were in great abundance. Potatoes, carrots, turnips, and cabbages, plenty; as well as feveral pleafant roots and greens which grew fpontaneous. Great variety of berries were found, and in fufficient quantities to yield fupplies of good drink. Befides which, we brewed fpruce-beer, and had a very plentiful ftock of tolerably good French brandy.

We were vifited by Verofhagin, the prieft of Paratounka, and his family; and I was very happy to meet with a number of the acquaintances of my countrymen in Captain Cook's Expedition. Nothing in nature could be more pleafant than the glow of friendfhip which animated their countenances with the livelieft expreffion of fincere regard, when they mentioned the names of King, Bligh, Philips, Webber, and others; names that will be

U   handed

handed down to posterity by tradition in a Kamtshatka song to their memory, with a chorus to the tune of *God save the King*; which is frequently sung in perfect harmony, particularly by the family of Veroshagin at Paratounka, by the different branches of which it was made. They deeply lamented the fate of Captain Clerke, whose tomb is now graced with an engraving on a sheet of copper, containing a copy of the superscription painted on the board, and suspended on the tree under which he lies buried ; with this addition only, " Erected by Perouse 1787, commander of the Expedition from France." Near this place is a half-decayed wooden cross, denoting the place of interment of the naturalist De Lisle de la Croyere, who died in Commodore Bering's Expedition.—[*See the annexed* ENGRAVING.]

We made frequent excursions and visits, and were well entertained with songs, dances, &c. of which I purpose giving an account hereafter, with a description of the country, and the customs and manners of the people.

Fine weather continued till the 16th November, when we had snow, and the appearance of winter setting in ; the thermometer being 2, 3, and 4 degrees below the freezing point of Reaumur.

It was now necessary to discover the most eligible place for building a vessel to accompany the Slava Russie. The only wood produced about the bay of Avatsha is birch ; but in the river Kamtshatka are considerable woods of larch, fir, and common pine. Captain Billings resolved upon visiting the Kamtshatka, in company with Captain Hall, and building in the Lower Town a cutter, as a consort to the ship

They

Drawn by W.Alexander.

Engraved by J.Powell.

View of Capt. Clerke's Tomb at St. Peter & St. Paul.

Published March 24th 1801, by Cadell & Davies, Strand.

They quitted the harbour of St. Peter and St. Paul, with Mr. Bakoff and Surgeon Robeck, on the 24th November; leaving Captain Saretſheff to ſuperintend. Doctor Merck, Mr. Vaconin (the draftſman), a bird ſtuffer, and neceſſary aſſiſtance, ſet out the 4th December on an excurſion, to deſcribe the hot ſprings, and collect natural curioſities. About Chriſtmas, Mr. Shmaleff, the commander of this diſtrict, arrived, and increaſed the harmony and good humour of our ſociety.

Parties were ſent to Bolſhoiretſk and Virchnoi, or the Upper Town, to prevent their quarters being heavy on the inhabitants; and materials were forwarded by ſmall quantities to the Lower Town for building the veſſel; for which purpoſe trees were felled.

We paſſed a pleaſant and agreeable winter in different excurſions to Bolſhoiretſk, &c. enjoying all the good things in Kamtſhatka, and perfect health. The froſt was generally 5° to 8°. The ſevereſt cold that we had did not exceed 18°, and laſted only a few hours. Snow was very deep.

CHAP.

CHAP. XII.

Receive information from Peterſburgh of an Enemy's Ship (Swediſh) having been ſent into theſe Seas to annoy the Ruſſian Fur-Trade.—Depart from Avatſka Bay.—Captain Billings cauſes his Inſtruc-tions to be read to the Officers, and declares his intention of ſteering to the North-weſt Coaſt of America.—Iſland of Amtſhitka.—Amli.—Oonalaſhka ; Dreſs, Manners, &c. of the Natives.—Ty-ranny exerciſed over them by the Ruſſian Hunters.

EARLY in the month of March 1790, we were all collected to-gether in the harbour of St. Peter and St. Paul, on the receipt of diſpatches from St. Peterſburg, confirming the Swediſh war, and informing us of a ſhip called the Mercury, mounting 16 guns, under the command of a Mr. Coxe, having been ſent into theſe ſeas by the Court of Sweden to annoy the Ruſſian fur-trade ; which it was to be our buſineſs to prevent.

Towards the end of April the harbour was clear of ice ; but the mountains were ſtill covered with ſnow, except ſuch parts as were oppoſed to the influence of the ſun, where vegetation began.

On the firſt of May all hands embarked, and the ſhip was hauled into the bay of Avatſha. The weather was clear and calm ; we obtained ſome wild garlic (Tſheromtſha), and obſerved the hawthorn and birch beginning to bud.

We

We had 16 brafs three-pounders mounted, and on the morning of the 2d took our powder on board. We obferved high water at the head of the inner harbour at full and change of the moon, at four hours and forty-three minutes: the greateſt rife fix feet.

Calms and contrary winds detained us till the 9th May, at four A. M.; when, a moderate breeze fpringing up from the north north-weft, we ſtowed away our boats, weighed anchor, and ſtood out of the bay, ſteering fouth fouth-eaſt, and keeping a middle courſe.

At eight A. M. the light-houſe bore north-weſt 15°, diſtant two leagues; our latitude 52° 49 north, longitude 158° 47' eaſt from Greenwich. From hence we took our departure; and at noon our latitude was 52° 46' 4", longitude 158° 54'.

At half-paſt feven in the evening, Avatſha Volcano bore north-weſt 35°, the light-houſe north-weſt 78°; and we ſhortly after loſt fight of land. The weather was raw and cold; our thermometer indicated at midnight one degree above the freezing point; and we had a moderate foutherly breeze, the fea running very high from the fouth-weſt.

On the 10th we had a moderate breeze from the fouth-weſt, with a rough fea and hazy weather; our courſe fouth-eaſt, which continued on the 11th. At noon our obſerved latitude was 51° 18' 6", longitude 161° 58'. During the day we faw feveral flights of ducks, numberlefs gulls, auks, and fea-parrots; a few grampufes and feals; and a plank very like the ſheathing of a ſhip. The afternoon was very hazy and cold. On the 12th, we had

* U 3

light

light airs from the fouth, and hazy weather. We faw during the day whales, feals, gulls, auks, &c. Our latitude, by account, 51° 5′, longitude 163° 50′.

In the morning of the 13th, Captain Billings had his inftructions read to the officers ; and told them, that his intention was to fteer for the iflands fouth of Alakfa, and to the north-weft coaft of America ; confidering the furveying of the chain of Aleutan iflands, fo inaccurately laid down on the charts, as too dangerous to be attempted with a fingle veffel during the foggy feafon.

We had a frefh breeze from the fouth, and our courfe was eaft north-eaft. We reckoned our latitude at noon 51° 5′, longitude 166° 30′. Towards evening, we had variable light airs, and foggy weather. Light winds from the eaft, and calms, continued during the 14th and 15th. The 16th, variable winds, and thick mifty weather. At noon faw ducks and fmall birds flying to the eaftward, and rock weeds floating ; latitude, by account, 50° 40′, longitude 169° 5′. In the afternoon a gentle breeze fprung up from the fouth and fouth-weft ; our courfe eaft by north : we faw feveral flights of land birds.

The 17th, variable winds and rain. At noon, obferved the latitude 51° 11′ 7″, longitude, by account, 170° 25′. At four P. M. a frefh breeze fprung up from the north north-weft with flying clouds at 4° 16′ 15″ apparent time ; longitude, by time-keeper, 171° 18′ eaft from Greenwich, latitude 51° 12′ 4″. The 18th, obferved the latitude at noon 50° 49′ 23″ ; faw great quantities of rock-weed floating, and numbers of porpoifes and birds. At four 50° 55′ P. M. apparent time,

5

time, longitude 173° 14', latitude 50° 49' 20''; variation of compafs 13° 10' eaft. The 19th, we had a frefh breeze north by weft; our courfe north-eaft by eaft; cloudy weather, with a hazy horizon. At noon, obferved the latitude 50° 44': the afternoon fqually, with rain. According to Captain Billings's cuftom, we went under an eafy fail every night, or laid-to. The 20th, at noon, we were in latitude 50° 27' 52'', longitude, by account, 175° 40'. This day and the 21ft we had variable light winds between the north and the eaft, with raw mifty weather, 3° above the freezing point, and a moderate fwell from the north-eaft. The 22d blew frefh from the eaft north-eaft; kept a northern courfe all day; mifty and hazy weather; night fqually, at times fnow: and on the 23d the wind veered to north north-weft, blowing frefh; the fea running very high; fteered north-eaft. At noon, faw great quantities of rock-weed floating, and land-birds flying to the north. Got a fight of the fun, which gave the latitude 51° 6' 43'', at 3° 41' 15'' apparent time, longitude, by time-keeper, 177° 57' 45'', latitude 51° 18'. At eight P. M. faw land north and north-eaft, on account of which, and approaching night, clofe-reefed top-fails, and handed all fmall fails; the wind veered to weft by fouth, we kept our fhip's head fouth by weft till day-light on the 24th, when we again ftood to the north-eaft. At 3° 30' A. M. faw high land; and, when we got well in with it, ranged along the fouth-fide. It was the ifland Amt-fhitka, the eaftern extremity of which bore at noon north 20° eaft, diftant about 12 miles. It commences to the weft with a low point of land, gradually rifing into moderate mountains, trending fouth 49°, eaft 25 miles; where it forms a head-land, from whence its direction is north 64°, eaft 14 miles: to the eaft and weft are detached rocky ifles. The mountains were covered with fnow, and no wood to be feen: our latitude, by account,

4   was

was 51° 18', longitude 179° 25'. Squally weather, with a hard gale from the south-west; the sea running very high, and the surf breaking violently on a reef of hidden rocks near the land; our course east. The 25th, wind more moderate; thick hazy weather; fresh gales, varying from south-west to south south-east; kept our course east and east north-east; latitude, by account, at noon 50° 46'; towards night slackened sail. The 26th, fresh easterly wind, which veered to east, north-east, north, and north-west by west, blowing hard with rain; the sea running very high. At eight P. M. the gale, with squalls, brought us under our courses, and we saw land in the haze; the western cape north, 14° west, distance about four leagues: the east extremity bore north. We supposed this to be the island of Adak; but, fearful of entangling the ship among islands so badly placed in our charts, it was thought proper to lay-to under mizen, main, and fore-sail, with the ship's head to the westward, to wait day-light. The 27th, at four A. M., wore ship, and stood to the north-east by east under the above sails, wind continuing north-west by west. At noon, our latitude observed was 51° 12' 57", corrected longitude 184° 55'; at three P. M. saw land, two mountains covered with snow, north 44° west, distant about 36 miles. At 3° 59' 50" P. M. apparent time, longitude, by time-keeper, 184° 35' 30", latitude 51° 18' 52"; at the same time saw land, a high mountain north-west 38°, distant about 30 miles; soon hid by the haze; and, on account of approaching night, slackened sail. The 28th, had a fresh gale at west south-west; our course north-east, the sea running very high, and cloudy weather. At ten A. M. saw land, which we supposed to be the island of Amli; and, to get a better sight of it, shaped our course north. At noon, the west cape of a bight bore north-west 21°; east cape north-west 15°, distant eight miles; latitude observed 51° 55' 23", longitude, by account,

187°

Plate V.

Alexander del.

Neagle Sculp.t

A Man and a Woman of Oonalashka.

Published March 2.d 1802, by Cadell & Davies. Strand.

187° 36'. The iſland of Amli, from the weſt point, trends 44 miles ſouth, 88° eaſt. At 4° 10' 25" P. M. apparent time, longitude, by time-keeper, 187° 12', latitude 51° 55' 9", the variation of two compaſſes gave the mean 17° 7' eaſt. The 29th, at noon, the latitude obſerved was 52° 23' 53"; correcled longitude 190° 14'; freſh breezes from the ſouth-weſt; our courſe north-eaſt. Till noon of the 30th, variable light airs; latitude obſerved 52° 34' 5", correcled longitude 191° 2'. In the afternoon, little wind from the north-weſt, ſteering north-eaſt at 5° 24' 25" P. M. apparent time; longitude by time-keeper 191° 2', latitude 52° 37' 7". We ſaw land to the north all night, and ſtood to the weſt ſouth-weſt. At day-light of the 31ſt reſumed our courſe north-eaſt. At 8° 23' 20" P. M. apparent time, our longitude was 191° 40', latitude 52° 40' 5". Towards noon rainy weather.

In the morning of the 1ſt June, at 4° 30' we roſe the iſland of Oonalaſhka in the north-eaſt. At 8° 5' 45" A. M. our longitude was 192° 41' 15", latitude 52° 51' 17", little wind at north north-weſt, which died away to a calm. At four P. M. had a gentle breeze from the north north-weſt again; and at 6° 26' 25" A. M. the time-keeper gave the longitude 193° 2' 15", latitude 52° 59'. On the 2d variable light airs and calms. At noon our latitude obſerved was 53° 3' 29", correcled longitude 193° 47'. We were now well in with the land of Oonalaſhka, which appeared every where high, formed of projecling promontories and inland high mountains. In the forenoon of the 3d June, numbers of the natives came alongſide. We threw our main-top-ſail to the maſt, and took them on board. At noon we obſerved the latitude 53° 45' 4". At four P. M. a Ruſſian hunter of Tſhirepanoff's company came alongſide in a baidar rowed by eight Alcutes. He had been along ſhore in ſearch of drift wood for firing,

X   with

with a number of Alcutes, fome of whom brought us a good fupply of Halibut. They conducted us into a bay which the Ruffians call Bobrovoi Guba, or the Bay of Otters, where we came
to anchor at eight P. M. oppofite the habitations of the natives.
We fent an officer to found, and hauled into the bay about 40
fathom from fhore.

Captain Billings landed with his aftronomical tent; Dr. Merck
went out on an excurfion for curiofities; and Captain Saretfheff,
with affiftants, was fent to furvey; while I employed myfelf in
getting the beft information that I could obtain of the inhabitants,
who with the people of Oomnak, call themfelves Cowghalingen.
This habitation they call Sidankin. It is on the fmall ifland Sithanak,
feven miles from north-eaft to fouth-weft, which is feparated from
Oonalafhka by ftraits of only a few fathom wide, and appears to
be the fouth-weft extremity of the ifland. It confifts of barren
mountains of a moderate height, compofed of hard ftone of a
glaffy nature, and generally of a greenifh hue: fome, however,
is black. Behind the huts is a lake of fome extent (evidently
fupplied by the melting fnow from the mountains), with a fmall
outlet or run into the fea. Here we took a fupply of frefh water,
which was not very good.

About five families refide here. The natives of Alakfa and all
the adjacent iflands they call Kagataiakung'n, or eaftern people:
the iflanders of Oone-agun (Tfhettiere Sopofhnoi) they call
Akohgun.

The people are of middle fize; of very dark brown and healthy
complexion; round face in general, fmall nofe, black eyes and
hair, the latter very ftrong and wiry. They have fcanty beards,

I

but

but very thick hair on the uper lip. The under lip is, in general, perforated, and fmall ornaments of bone or beads inferted; as is alfo the feptum of the nofe. Women have the chin punctured in fine lines rayed from the centre of the lip, and covering the whole of the chin. The arms and cheeks of fome are alfo punctured. They are very clean in their perfons; and the men very active in their fmall baidars. The women are chubby, rather pretty, and very kind.

They formerly wore a drefs of fea-otter fkins, but not fince the Ruffians have had any intercourfe with them." At prefent they wear what they can get; the women, a park of kotik, or urfine feal, with he hair outward. This is made like a carter's frock, but without a flit on the breaft, and with a round upright collar, about three inches high, made very ftiff, and ornamented with fmall beads fewn on in a very pretty manner. Slips of lea-ther are fewn to the feams of this drefs, and hang down about 20 inches long, ornamented with the bill of the fea-parrot, and beads. A flip of leather three or four inches broad hangs down before from the top of the collar, covered fancifully with different co-loured glafs-beads, and taffels at the ends: a fimilar flip hangs down the back. Bracelets of black feal fkin are worn round their wrifts about half an inch broad, and fimilar ones round their ankles, for they go barefooted; and this is all their drefs. Their ornaments are rings on the fingers, ear-rings, beads and bones fufpended from the feptum of the nofe, and bones in the per-forated holes in the under lip. Their cheeks, chin, and arms, are punctured in a very neat manner. When they go a-walking on the rocky beach, they wear an aukward kind of boot, made of the throat of the fea-lion, foled with thick feal-fkin, which they line with dry grafs. The men wear a park of birds' fkin, fome-

times

times the feathers outward, and sometimes inward.   The skin side
is dyed red, and ornamented with slips of leather hanging down
a considerable length; the seams covered with thin slips of skin,
very elegantly embroidered with white deer's hair, goat's hair, and
the sinews of sea animals, dyed of different colours.   They also
wear tight pantaloons of white leather, and boots as described to
be worn by the women at times : the men wear them when they
go on foot; but in their baidars or their huts they are without
either pantaloons or boots.   The men have their hair cut short ;
the women wear theirs short before, combed over the forehead,
and tied in a club on the top of the back part of the head.   In
wet weather, or when out at sea, they wear a camley ; which is a
dress made in the shape of the other, but formed of the intestines
of sea animals ; the bladder of the halibut, or the skin off the
tongue of a whale.   It has a hood to cover the head, and ties
close round the neck and wrists ; so that no water can penetrate :
it is nearly transparent, and looks pretty.   The men wear a
wooden bonnet, ornamented with the whiskers of the sea-lion,
and with beads, which make very pretty nodding plumes ; and
this serves to fasten the hood of their camley to the head.   The
women's park is called tshoktakuk, the men's iash ; the boots,
ooleegich.   Both men and women are very fond of amber for
ornaments, as also of a thin shelly substance formed by worms in
wood, about two inches long, thin, tapering, and hollow.

Their instruments and utensils are all made with amazing
beauty, and the exactest symmetry ; the needles with which they
sew their clothes and embroider are made of the wing-bone of
the gull, with a very nice cut round the thicker end, instead of an
eye, to which they tie the thread so skilfully, that it follows the
needle without any obstruction.   Thread they make of the
                                                                  sinews

finews of the feal, and of all fizes, from the finenefs of a hair to the ftrength of a moderate cord, both twifted and plaited; the plaited cords of their darts, to which they tie the gut of the feal blown out to ferve as a float, are very beautifully ornamented with red downy feathers, and goat's hair; as are alfo the different ftrings with which they faften the wrifts and other parts of their cloath- ing, &c.

Their darts are adapted with the greateft judgment to the dif- ferent objects of the chafe; for animals, a fingle barbed point; for birds, they are with three points of light bone, fpread and barbed; for feals, &c. they ufe a falfe point, inferted in a focket at the end of the dart, which parts on the leaft effort of the ani- mal to dive, remaining in its body. A ftring of confiderable length is faftened to this barbed point, and twifted round the wooden part of the dart; this ferves as a float to direct them to the feal, which, having the ftick to drag after it, foon tires, and becomes an eafy prey. It, however, requires fkill to humour it, perhaps equal to our angling. The boards ufed in throwing thefe darts are equally judicious, and enable the natives to caft them with great exactnefs to a confiderable diftance.

The baidars, or boats, of Oonalafhka, are infinitely fuperior to thofe of any other ifland. If perfect fymmetry, fmoothnefs, and proportion, conftitute beauty, they are beautiful; to me they ap- peared fo beyond any thing that I ever beheld. I have feen fome of them as tranfparent as oiled paper, through which you could trace every formation of the infide, and the manner of the natives' fitting in it; whofe light drefs, painted and plumed bonnet, toge- ther with his perfect eafe and activity, added infinitely to its ele- gance. Their firft appearance ftruck me with amazement beyond expreffion.

expreſſion. We were in the offings, eight miles from ſhore, when they came about us. There was little wind, but a great ſwell of the ſea: ſome we took on board with their boats; others continued rowing about the ſhip. Nearer in with the land we had a ſtrong rippling current in our favour, at the rate of three miles and a half, the ſea breaking violently over the ſhoals, and on the rocks. The natives, obſerving our aſtoniſhment at their agility and ſkill, paddled in among the breakers, which reached to their breaſts, and carried the baidars quite under water; ſporting about more like amphibious animals than human beings. It immediately brought to my recollection, in a very forcible light, Shakeſpeare's expreſſion —

> " He trod the water,
> " Whoſe enmity he flung aſide, and breaſted
> " The ſurge moſt ſwoln that met him."

Theſe baidars are built in the following manner: A keel eighteen feet long, four inches thick on the top, not three inches deep, and two inches, or ſomewhat leſs, at the bottom. Two upper frames, one on each ſide, about an inch and a half ſquare, and ſixteen feet long, join to a ſharp flat board at the head, and are about ſixteen inches ſhorter than the ſtern, joined by a thwart which keeps them about twelve inches aſunder. Two ſimilar frames near the bottom of the boat, ſix inches below the upper ones, about one inch ſquare. Round ſticks, thin, and about ſix inches diſtant from each other, are tied to theſe frames, and form the ſides; for the top thwarts, very ſtrong ſticks, and nearly as thick as the upper frames, curved ſo as to raiſe the middle of the boat about two inches higher than the ſides. There are thirteen of theſe thwarts or beams: ſeven feet from the ſtern is one of them; twenty inches nearer the head is another; a hoop about

two

two inches high is faftened between them, for the rower to fit in. This is made ftrong, and grooved to faften an open fkin to, which they tie round their body, and it prevents any water getting into the boat, although it were funk. This frame is covered with the fkin of the fea lion, drawn and fewn over it like a cafe. The whole is fo extremely light, even when fodden with water, that it may be carried with eafe in one hand. The head of the boat is double the lower part, fharp, and the upper part flat, refembling the open mouth of a fifh, but contrived thus to keep the head from finking too deep in the water; and they tie a ftick from one to the other to prevent its entangling with the fea weeds. They row with eafe, in a fea moderately fmooth, about ten miles in the hour, and they keep the fea in a frefh gale of wind. The paddles that they ufe are double, feven or eight feet long, and made equally neat with the other articles.

The women plait very neat ftraw mats and bafkets; the former ferve for curtains, feats, beds, &c.; the latter to contain their work and other implements. Their trinkets and coftly ornaments are kept in fmall wooden boxes with draw lids.

I obferved in all the huts a bafket containing two large pieces of quartz, a large piece of native fulphur, and fome dry grafs or mofs. This ferves them in kindling fires; for which purpofe they rub the native fulphur on the ftones over the dry grafs, ftrewed lightly with a few feathers in the top where the fulphur falls; then they ftrike the two ftones one againft the other; the fine particles of fulphur immediately blaze like a flafh of lightning, and, communicating with the ftraw, fets the whole in a flame.

Their

Their only mufic is the tambour, to the beat of which the women dance. Their holidays, which are kept in the fpring and autumn, are fpent in dancing and eating. In the fpring holidays, they wear mafks, neatly carved and fancifully ornamented. I believe that this conftitutes fome religious rite, which, however, I could not perfuade them to explain: I attribute this to the ex-traordinary and fuperftitious zeal of our illiterate and more favage prieft *, who, upon hearing that fome of our gentlemen had feen a cave in their walks, where many carved mafks were depofited, went and burnt them all. Not fatisfied with this, he threatened the natives for worfhiping idols, and I believe I may fay *forced* many to be chriftened by him, without being able to affign to them any other reafon than that they might now worfhip the Trinity, pray to St. Nicholas and a crofs which was hung about their necks, and that they would obtain whatever they afked for; adding, that they muft renounce the devil and all his works, to fecure them eternal happinefs. It appeared to me that they re-garded this as an infult; be that as it may, however, they were not pleafed, but had not power to refent.

They have no marriage ceremony among them, but purchafe of father and mother as many girls as they can keep; and, if they repent of their bargain, the girl is returned, and a part of the purchafe given back. They formerly ufed to keep objects of unnatural affection, and drefs thefe boys like women.

* I have called the prieft more than favage, and fhall relate a circumftance that hap-pened in proof. While he was travelling from Yakutfk to Ochotfk, he loft fome provifion on the road. On a mere fuppofition that his two Tartar guides had taken it, he tied each of them up by an arm to a tree, and had them flogged to fuch a degree, that one of them died, and the other never recovered the ufe of his arm: it was afterwards known, that fome runaway exiles hid in the woods were the thieves. The prieft faid, *there was no harm done;* *they were not Chriftians.*

At

At births alſo, no ceremony is uſed, except waſhing the infant.

They pay reſpect, however, to the memory of the dead ; for
they embalm the bodies of the men with dried moſs and graſs ;
bury them in their beſt attire, in a ſitting poſture, in a ſtrong box,
with their darts and inſtruments ; and decorate the tomb with
various coloured mats, embroidery, and paintings. With women,
indeed, they uſe leſs ceremony. A mother will keep a dead child
thus embalmed in their hut for ſome months, conſtantly wiping
it dry ; and they bury it when it begins to ſmell, or when they
get reconciled to parting with it.

They dry ſalmon, cod, and halibut, for a winter's ſupply, and
collect edible roots : this, however, is not for themſelves, but for
ſuch Ruſſian hunters as may chance to viſit them. At this time
there are twelve Ruſſians and one Kamtſhadal, of Tſhirepanoff's
company of hunters on the iſland. They have lived here eight
years, but are going this year back to Ochotſk. Theſe people
lord it over the inhabitants with more deſpotiſm than generally
falls to the lot of princes ; keeping the iſlanders in a ſtate of abject ſlavery ; ſending parties of them out on the chaſe, and to
their veſſel, which now lies in the Straits of Alakſa ; ſelecting
ſuch women as they like beſt, and as many as they chooſe. They
ſeem to me to have no deſire to leave this place, where they enjoy that indolence ſo pleaſing to their minds ; for, by changing
of places, they change ſituations, and become themſelves as much
the ſlaves of power, as the poor natives are to them.

I obſerved, in croſſing the mountains, piles of ſtones. Theſe
are not burying-places, as has been ſuppoſed, but ſerve as beacons

Y   to

to guide them in foggy and ſnowy weather from one dwelling to the other; and every perſon paſſing adds one to each heap. The only obſervations that we made on ſhore proved our time-keeper ſtill going as when we left Kamtſhatka; our latitude 53° 56', longitude 194° 20'; variation of the compaſs 19° 35' eaſt.

CHAP.

CHAP. XIII.

Depart from Oonalashka.—See the Island of Sannach.—The Shu-
magins.—Aleutes oppressed by the Russian Hunters.—Islands
of Evdokeeff.—Come to anchor in the Harbour of Kadiak.—Par-
ticulars respecting that Island and its Inhabitants.

ON the 13th June, at eight A. M , having taken in a supply of
fresh water, and ballast, we weighed anchor, with a gentle breeze
north-west ; but, finding that we could not weather the rocks off
the eastern cape, again came to anchor near our old station, where
we remained till the evening of the 17th, when, the wind shift-
ing more to the north, we weighed, and got out.   Our soundings
in the bay were 17, 16, 15 fathoms ; suddenly deepening so, that
we got no bottom close in with land with 100 fathoms line.   At
midnight hoisted in our boats, and proceeded with light airs and
foggy weather.   The 18th, at 10 A. M., the volcano on the island
of Akutan bore north-west 81°.   Three conical mountains on the
island of Oonimak bore, the first, north-west 18° ; the second,
north-west 2° ; the third, north-east 12°.   At noon, our distance
made was 41 miles south-east 86°; latitude observed 53° 52' 6",
longitude 194° 43'.   At one P. M. the volcano in Oonalashka
south-west 77° ; west cape of Akutan south-west 86° ; the first
mountain in Oonimak, called by the natives Koogidan Kaigut-
shin, north-west 12° ; the volcano called Agaiedan, north-
east 8° ; the third mountain, Khaiginak, north-east 15°.   At

Y 2   4° 30' 55"

4° 30' 55" apparent time, the longitude proved, by time-keeper,

keeper, - - -	195°	35'	15"
Latitude -	53°	58'	6"
Variation of the compass	0°	19'	40"

The evening was very foggy, with rain, light airs from the weftward, and calms. At midnight we got foundings with 60 fathom line; a muddy and black fandy bottom.

The morning of the 19th was very foggy, with variable light airs from fouth fouth-weft to fouth fouth-eaft. We faw a number of the kotic, or urfine feals, fporting about the veffel; alfo one fea-otter. At eight o'clock we had a gentle breeze from the fouth-eaft, hazy, and rain; foundings 30 fathoms. We were well in with the ifland of Oonimak, upon which the fog refted; it trends from the weftern extremity fouth-eaft 63°, 18 miles; to the northern cape 19 miles in a direction north-eaft 62°. The land is high, broken, and rugged, and there are three very confpicuous mountains upon it. The fummit of the firft is very irregular; the fecond is a perfect cone towering to an immenfe height, and difcharging a confiderable body of fmoke from its fummit; the third (Khaginak) has its fummit apparently rent and broken, covered with fnow, and towering above the fog which covered the middle of the land. On the lower parts of them, and in the vallies, no fhrub nor bufh was to be feen: our latitude at noon was, by reckoning, 54° 25', longitude 196° 6'. In the afternoon we had a frefh breeze from the fouth-eaft by eaft; our courfe north-eaft by eaft; keeping the lead going; foundings pretty regular for 30 to 45 fathom. At eight P. M. we tacked and ftood fouth by weft and fouth away from the land.

                                                          The

The 20th, at three A. M. we again tacked for the iflands, and shaped our courfe to the eaftward; our foundings varying from 46 to 33 fathom, fmall ftones, fhells, and fand. It blew frefh, and was hazy with rain. At half paft four in the morning, the weather cleared up a little, and difcovered the ifland of Sannach, with rocks and breakers over a reef, about a mile right ahead of the fhip, fo that we had but juft time to get clear of them. This ifland is inhabited by a few Aleutan families; and in the middle of it are three confiderable mountains, joining together. The eaft and weftern extremities are low land, and appear verdant, but without wood. It is furrounded by a reef of rocks, fome above water, and the furf breaking violently over others. At noon our latitude was 54° 22', longitude 197° 37', and our diftance from land 12 miles. The eaft cape bore fouth 55° weft; the weftern cape fouth-weft 57° 30'. I compute the length of the ifland at 15 miles. Cape Alakfa lies nearly north of the mountains on this ifland; the diftance about 38 miles.

Shortly after feeing this place (Cook's Halibuts Ifland) we rofe a number of fmaller, forming the group called Shumagins, from their firft difcoverer,—a failor in Bering's expedition. They extended from north-weft to north-eaft, as far as the eye could reach. At one in the morning we were pretty clofe in with the moft remarkable of them, called by the natives Animok, and by the Ruffians Olenoi. This is very high and bluff; the others are lefs elevated. This is in latitude 54° 44', longitude 198°, and about ten leagues from Alakfa. All thefe iflands are furrounded by rocks, fome above water, and others only difcernible by the breakers. We could not get a good view of the main land, on account of the thick weather; but now and then faw the tops of tremendous mountains covered with fnow.

We

We had a frefh breeze from the weft, failing under clofe-reefed topfails at about fix miles; north-eaft and north north-eaft; hazy, with a rough fea. Towards night the wind fell fcant. On account of the immenfe number of iflands, we ftood to the weft fouth-weft and fouth-weft, with the intent of ftanding in again at day-light to view them.

Early in the morning of the 21ft, we had a gentle breeze from north north-weft, with pretty clear weather. At 4° we obferved cape Alakfa north-weft 68°, at 8° 31′ 45″; our time-keeper gave the longitude 199° 32′ 45″, latitude 55° 3′ 54″. The iflands Nagai and Kagai being right a-head, we wifhed to go between them; but the breakers induced us to give up this idea, and vary our courfe from eaft north-eaft to eaft by fouth, with a wefterly breeze, going at four knots. At 5° 17′ 20″ P. M. longitude by time-keeper 200° 35′, latitude 54° 54′ 24″. At half paft feven in the evening, we obferved a three-feated baidar, and five fmaller ones, near the fhip. When they came alongfide we hoifted the rowers and their baidar on board, and learned that they were out on the chafe for fea lions and feals; that their company confifted of more than one hundred Aleutes, under the direction of the Ruffian in the three-feated baidar; and that they were hunting for Panofsky's veffel, now lying off Alakfa. We laid-to about three hours with our main topfail to the maft; and at day-break of the 22d fent them off. The Aleutes left us with reluctance; and complained bitterly of the treatment that they met with, and of being compelled to ferve for years without receiving any recompence. We, however, had interpreters on board, and could not keep them. At noon our obferved latitude was 55° 9′ 27″, variable light airs. At 5° 7′ 45″ P. M. our time-keeper gave the longitude 201° 59′, latitude 55° 11′ 20″.

All

All the 23d we had variable light airs and calms, with foggy and hazy weather.

The group of Shumagin iflands are very clofe together, almoft innumerable, and extend from Alakfa fouth 15 or 16 leagues, and from eaft to weft about 60 leagues; all of them high and barren, exhibiting a great fimilarity in their appearance, though of various forms and fizes. Neither tree nor fhrub is to be feen upon them. The low places appear green; but a brownifh hue diftinguifhes all the higher mountains, except fuch as are covered with fnow. Some project into the fea in rugged cliffs; fome are fharp capes, and often terminate in bluff heads. There appear fome convenient coves; but it would be hazardous to enter them, on account of the detached and funken rocks that are fcattered about: nor are there any navigable ftraits between them; but they muft be paffed either to the fouth, or clofe in with Alakfa. Thefe iflands are generally frequented by fome company of hunters for feals, fea lions, and birds. Sea otters are very feldom found hereabout; but whales are very numerous, as are feals of different kinds, &c. We fent our jagers out in the boats during the calm to fhoot birds, fome of which we found very good eating.

Variable winds and hazy weather continued till the 24th at noon, when we got a fight of the fun, and our latitude proved 55° 41' 7', longitude 201° 43'. About half paft twelve a gentle breeze fprung up from the fouth; and, failing north by eaft about two knots, at one we faw land ahead. At feven in the afternoon we rofe more land, bearing north-eaft and fouth-eaft; foundings 37 fathoms, fine fand. At midnight bore away to the fouth-weft about two hours, when we again ftood in for land, which is

called

called by the Ruffians the iflands of Evdokeeff (the largeft, Sime-
dan), which bore, the 25th at noon, fouth-eaft, diftant about 10
miles; our latitude obferved, 56° 10' 40", longitude corrected,
202° 55.' We fent out a jager, with two Aleutes, to fhoot birds.
At two we had variable light winds and cloudy weather. Three
baidars came along-fide, with a Ruffian hunter from Shelikoff's
Eftablifhment at Kadiak, having about 200 natives in company,
in chafe of fea-lions, kotic, and other feals, birds, &c. At three
P. M. being about three miles from the iflands, Captain Billings
went on fhore with the naturalift. It was almoft calm ; and at
five P. M. he returned, defcribing the iflands as complete rocks of
coarfe granite, with a few ftunted vegetables growing on the low
parts. Our three hunters (who went out in a three-feated baidar
purchafed at Oonalafhka), not returning, we fired fignal guns.
At fix P. M. the natives, &c. left us, and the Ruffians promifed to
feek our men and fend them off. We lay-to all night, with our
main-top-fail to the maft, and kept a lanthorn at the mizen peak.
At four o'clock in the morning of the 26th, a gentle breeze fprung
up from the fouth-eaft, and we kept plying off the iflands, with
very hazy weather. At fix the men came on board, telling us,
that they had loft fight of the fhip in the haze, and only faw her
this morning at day-light, when they put off. At 7° 50' 10" we
got a fight of the fun ; longitude, by time-keeper, 203° 20', latitude 56° 15' 39". At noon, our obferved latitude was 56° 20'
24" ; the afternoon cloudy, with a moderate breeze from fouth-
eaft, fteering north-eaft by eaft. At two P. M. we faw the ifland
Okamok in the hazy horizon, bearing fouth-eaft 56°, diftant about
fix leagues ; and at eight P. M. it fell little wind and rain.

June 27th, at five A. M., we faw the lofty mountains on the
ifland of Kadiak ; and at eight P. M. faw the low iflands Tooge-

dach

dach and Sichtunach, diftinguifhed in Captain Cook's chart by the name of Trinity ifland.

The 28th, at break of day, we got in pretty well with the ifland of Kadiak, keeping an eafterly courfe with variable light winds between the north and weft, and a clear fky. At five A. M. numbers of the natives came off in their canoes, of whom the greater part were taken on board and ferved as pilots, defcribing with great accuracy the depth of every opening.

The fouth extremity of this ifland forms a low point of land called by Captain Cook Trinity Cape. It runs out narrowing from mountains of a moderate height. Off this cape, at the diftance of one mile, is the ifland Anayachtalak, which Shelikoff names Egichtalik, about three miles and a half north of Trinity Ifland. The ftraits have foundings from 36 to 16 fathom, over a bottom of coral and fhells. Toogidach is low and barren; Sichtunach lies to the eaft of it three miles. This is low in the middle, with a fmall bay, but both the extremities are high land. The weftern part of Kadiak, though mountainous, is interfperfed with vallies, which produce only a few low fhrubs, and appear at a fmall diftance like inlets; but a barrier of furrounding rocks forbids the approach of any veffel. In the afternoon we had gentle gales from the fouth-weft and weft; our courfe was north-eaft, and north-eaft by north. At 4° 45' 50" apparent time, our longitude proved, by the time-piece, 205° 50' 30", latitude 56° 49' 8"; variation 27° eaft. When the north-eaft bluff cape of Sichtunach bore fouth-weft 49°, diftant two miles and a half, the eaftern point of Anayachtalak was north-weft 79°, by compafs, diftant about two miles. Having paffed thefe ftraits at 10 o'clock in the evening, with a wefterly breeze and fqualls, we got clofe

Z　　　　　　　　　　　　　　in

in with a fmall ifland called by the natives Nafikan, 200 fathoms
from Kadiak. We paffed thefe ftraits with 24 and 25 fathoms
water over a rocky bottom, and foon after fhaped our courfe
north north-weft for the bay. Nafikan is a remarkable ifland,
two miles long, and one wide; confifting of two round-topt mountains, which caufed Captain Cook to name it Two-headed Point.
Another projecting promontory, three miles from Two-headed
Point, in a direction eaft by north, he called Cape Barnabas.
This is the fouthern cape of the ifland Kunakan, or Kukan, and
bears fouth-eaft of the bay, Treck Svatiteley, in which Shelikoff has
his Eftablifhment: between thefe two iflands are the ftraits that
lead to the bay, which is about a mile wide at the mouth, where
foundings increafe to 50, 70, and 75 fathoms. On entering the
bay the 29th at day-break, with a moderate breeze from the weft
and weft by fouth, we could not get foundings with 100 and 150
fathom line. The fhores are fteep and rocky; fome detached
rocks run out a great way, and are nearly hidden at high water.
We found great difficulty in getting into the harbour, owing to
contrary winds; and the great depth prevented our fending out
a kedge. We therefore plyed to windward till half-paft three in
the afternoon, when we got above the harbour clofe in with the
weftern fhore, where we fent a tow-line with all the hands that
we could fpare, who, affifted by the natives, hauled the fhip into
the fmall harbour, and we came to anchor in a muddy bottom,
eight fathom, at fix P. M. Here we made the following obfervations:

This and the nearer iflands are inhabited by about 1300 grown
males, and 1200 youths, with about the fame number of females,
according to the regifter kept by Shelikoff's Eftablifhment, now
under the direction of Yefftrat Ivanitfh Delareff, a Greek; who
                                                    informed

informed me, that he had now out on the chafe, for the benefit of the Company, upwards of 600 double baidars of the natives, containing each two or three men. Thefe are divided into about fix parties, each under the direction of a fingle Peredoffhik, or Ruffian leader. Befide thefe, fmall parties are fent out daily to fifh for halibut, cod, &c. Females are employed in curing and drying fifh ; in digging, wafhing, and drying edible roots ; in collecting ufeful plants, berries, &c. ; and in making the dreffes of the natives, as alfo for the Ruffians. About two hundred of the daughters of the chiefs are kept at the Ruffian habitations near our anchoring place, as hoftages for the obedience of the natives ; and, as far as I could learn, they are perfectly well fatisfied with the treatment they meet with. The males are lefs fatisfied ; and, at the firft arrival of the Ruffians, feemed inclined to oppofe their refiding on the ifland ; but Shelikoff, furprifing their women collecting of berries, carried them prifoners to his habitation, and kept them as hoftages for the peaceable behaviour of the men, only returning wives for daughters, and the younger children of the chiefs. Every confiderable habitation of the natives had large baidars capable of containing forty or fifty men. Thefe were all purchafed by Shelikoff ; and the natives are now in poffeffion only of fmall canoes, none of which carry more than three. They feem reconciled to the rules introduced by the prefent chief of the company, Delareff, who governs with the ftricteft juftice, as well natives as Ruffians, and has eftablifhed a fchool, where the young natives are taught the Ruffian language, reading, and writing. He allows a certain number of the hoftages to vifit their relations for a ftipulated time ; thefe returning, others are allowed to go ; and, upon application of any one for his child's abfence, it is not refufed. The whole number of hoftages is about three hundred.

Z 2

The

The males are employed in the chafe in rotation, as are alfo the females: I mean, for the benefit of the community; for they lay in an amazing ftock of provifions, roots, berries, &c. to be fufficient for a winter's .fupply for the whole ifland, natives as well as Ruffians; a circumftance which feems, more than any thing elfe, to convince the favages that the Ruffians are not their abfolute enemies; for Delareff fays, that they never laid in a fup-ply of food for the winter till the Ruffians taught them; but, in bad weather, were obliged to collect cockles, mufcles, and other fhell-fifh, or refufe of the fea.

Luxuries, fuch as tobacco, beads, linen, fhirts, and nankeen dreffes, they pay for in particular. I obferved, that fuch of the parties as were fuccefsful in procuring rich fkins, received a fti-pulated payment; for each fea-otter, a ftring of beads about four feet long; for other furs in proportion; and that only food and the fkins of feals were the property of the community, of which the natives certainly enjoy the greater fhare, being by far the more numerous; and the fkins of feals are chiefly ufed by the natives to mend their baidars, and make new ones; in the latter cafe, they are purchafed for furs, foxes, marmot, otters, &c. or by fervice.

This Eftablifhment confifts of about fifty Ruffians, includ-ing officers of the company, and Sturman Ifmailoff, who is here, on the part of government, to collect tribute: this is the fame Ruffian officer that was feen, by Captain Cook's Expedition, at Oonalafhka, in the year 1778. He was one of the affociates of Benyowfky's confpiracy (by his own account forced away); but Benyowfky only carried him to one of the Kuril iflands, where

he

he flogged him and put him on fhore, with feveral others that were difaffected.

The buildings confift of five houfes after the Ruffian fafhion. Barracks laid out in different apartments, fomewhat like the boxes at a coffeehoufe, on either fide, with different offices : An office of appeal to fettle difputes, levy fines, and punifh offenders by a regular trial ; here Delareff prefides ; and I believe that few courts of juftice pafs a fentence with more impartiality : An office of receival and delivery, both for the company and for tribute : The commiffaries' department, for the diftribution of the regulated portions of provifion : Counting-houfe, &c. : all in this building, at one end of which is Delareff's habitation. Another building contains the hoftages. Befide which, there are ftorehoufes, warehoufes, &c. rope-walk, fmithy, carpenters' fhop, and cooperage.

Two veffels (galliots) of about 80 tons each are now here, quite unrigged, and hauled on a low fcaffold near the water's edge. Thefe are armed and well guarded, and ferve for the protection of the place. Several of the Ruffians have their wives with them, and keep gardens of cabbages and potatoes, four cows and twelve goats. Delareff is of opinion, that corn will grow near the eftablifhment which they are about forming in Cook's river.

One of the Ruffian officers, who has cohabited with a female native fome years, and has had feveral children by her, applied to our prieft to chriften her in form, and then join them together in the holy bands of matrimony ; which was done. She is a handfome woman, but punctured on the chin, and her under lip is perforated. Her houfe was extremely clean, as were alfo her children, and the latter apparently very healthy. She was dreffed in

14   the

the Siberian fashion, and seemed perfect mistress of Russian economy. I dined with them, and was very well satisfied with the treatment that I met with.

It was matter of amazement to me, while in Irkutsk, Yakutsk, and Ochotsk, to hear the very high wages given by Shelikoff to his common sailors; being from 600 to 1000 silver rubles yearly: their engagement, however, obliges them to purchase all their necessaries and luxuries of the Company at the market price. Here is only one market, which is the Company's stock; and the prices of articles are as follow: Brandy, one ruble per glass; tobacco, 50 rubles per lb. and sometimes more; a shirt, made of Russian coarse check, something resembling buntine in the looseness of the thread, 10 rubles; boot legs, without soles, 15 rubles and upwards; and every thing in proportion: so that their expences (they not being allowed to trade) exceed their salaries. Some of the men bitterly complained of this; but they laid nothing to the charge of Delareff: on the contrary, every one, native and Russian, spoke highly in his favour, and acknowledged several indulgencies received at his hands.

Shelikoff has called this island Kichtak, as the original name of it; in which, however, he is mistaken; for Kichtak, or Kightak, is merely an island; they call the Trinity Island Kightak Sichtunak; this, Kightak Kadiak; and, to my astonishment, one of them called Alaksa a Kightak, or island; and affirmed, that there were straits three days' row to the north of Kadiak. I made it my business to ascertain this, if possible; but had not an opportunity of learning any more, than that a river from a lake fell into the sea west of Kadiak, and that they carried their boats over a low mountain to an inlet, which communicated with Bristol Bay.

Bay. This was known to the Ruffian hunters and feveral natives, who, in confirmation, faid, that they obtained the tufks of the Walrufs, or Morzſh, from the oppofite fhore of Alakſa: their beſt fpears were pointed with them.

The natives call themfelves Soo-oo-ît, and their magicians Kanghémeut. I could not obtain any name from them for the Almighty; although they fay, that there is a fuperior being who has the command of all the fpirits; and that the wrath of thefe fpirits is only to be appeafed by offerings, and in fome cafes their flaves are facrificed, but very feldom; for all the prifoners that they take in their wars (which are almoft perpetual, one tribe againſt another) become flaves, and are fubject to ill treatment, particularly from the women. The female prifoners are all flaves, and fold from one tribe to another for trinkets, inftruments, &c. Not only their prifoners, however, are their labourers or flaves, but orphans become the property of thofe who bring them up, and are frequently redeemed by the relations of the parents; efpecially fuch as were inhabitants of other iflands.

The dwellings of the natives differ from thofe of Oonalafhka. They are but very little funk in the ground, and have a door fronting the eaft, made of a framed feal fkin; a fire-place in the middle; a hole over it, through the roof of the houfe, which ferves at once for the difcharge of fmoke, and the admiffion of light. The fides, partitioned off for fleeping and fitting places, are covered with grafs mats, much coarfer made than thofe of Oonalafhka. Each hut, or dwelling, has a fmall apartment attached to it, which ferves for a vapour bath; ftones are heated in the open air, and carried into thefe places, where the heat is

increafed

increafed to any degree by the fteam from water which is poured upon them.

The cuftoms of thefe favages are nearly allied to thofe of the Oonalafhkans. They have the fame kind of inftruments, darts, and boats, or baidars; but much worfe made; nor are they fo active upon the water. Their dances are proper tournaments, with a knife or lance in the right hand, and a rattle in the left; the rattle is made of a number of thin hoops, one in the other, covered with white feathers, and having the red bills of the fea-parrot fufpended on very fhort threads; which, being fhaken, ftrike together, and make a very confiderable noife: their mufic is the tambourine, and their fongs are warlike. They frequent-ly are much hurt, but never lofe their temper in confequence of it. In thefe dances they ufe mafks, or paint their faces very fantafti-cally. The dances of the women are only jumping to and fro upon their toes, with a blown bladder in their hand, which they throw at any one whom they wifh to relieve, and who always accepts the challenge.

The firft character, is the athletic and fkilful warrior; the fe-cond, the fleet and expert hunter; the former enjoys his prifon-ers and the booty of his enemy; the latter has his wives, la-bourers, and flaves by purchafe, and the ability that he poffeffes to maintain them. The moft favoured of women is fhe who has the greateft number of children. The women feem very fond of their offspring; dreading the effects of war, and the dangers of the chafe; fome of them bring up their males in a very effeminate manner, and are happy to fee them taken by the chiefs, to gratify their unnatural defires. Such youths are dreffed like women, and taught all their domeftic duties.

There

Plate VI.

Alexander del.

R.H. Cromek Sculp.

A Man of Kadiak

Published March 2ᵈ. 1802, by Cadell & Davies, Strand.

There is no ceremony in marriage: the ability to fupport women gives authority to take them, with their confent; in which cafe, the couple are conducted by the relatives of the girl to the vapour bath, which is heated, and they are left together; but fome prefent is generally made to the girl's father and mother. I inquired whether they lent their wives to one another? They told me, No; unlefs they were barren, and defired it; if they then had a child, they became the property of its father.

No other ceremony is obferved at births, than wafhing the child, and giving it a name.

The dead body of a chief is embalmed with mofs, and buried. The moft confidential of his labourers are facrificed and buried with him; alfo his inftruments of war or the chafe, and fome food. Numbers of the natives are baptized; but Delareff, the director of the Company, would not allow our prieft to compel any to become Chriftians; he, however, affifted him in perfuading as many as he could. Such as were at the fchool eftablifhed, willingly embraced the Greek religion, as did alfo numbers of the women.

The dreffes of the natives are the fame as at Oonalafhka, but worfe made; they are open about the neck, and have but very few ornaments. They are extremely fond of blue beads and amber, and carry on a trade with the natives of the neighbourhood of Cook's River, where they purchafe their baidars and canoes for trinkets, provifions, and oils of whales and feals. They ufe darts and lances headed with flate, with which they kill the fea animals. They alfo ufe poifon to their arrows, and the Aconite is the drug adopted for this purpofe. Selecting the roots of fuch

A a

plants

plants as grow alone, thefe roots are dried and pounded, or grated; water is then poured upon them, and they are kept in a warm place till fermented: when in this ftate, the men anoint the points of their arrows, or lances, which makes the wound that may be inflicted mortal.

They treat their vifitors, upon firft entering their dwellings, with a cup of cold clean water. When they have refted a while from the fatigue of rowing or walking, they put before them whale's flefh, the meat of fea lion, fifh, berries mixed with oil, and boiled farana, alfo mixed with fifh oil; and it is expected that the gueft fhall eat all that is fet before him. In the meantime their bath is heated, and the gueft is conducted into it, where he receives a bowl of the melted fat of feals or bears, to drink. The more the gueft eats and drinks, the greater honour is done to the hoft; but if he cannot eat all that is put before him, he muft take the remains away with him.

They begin their chace in February on the fouth fide of Kadiak; for the kotic it continues all March; in April they depart from Kadiak to the neighbouring iflands for fea otters, which are in the greateft perfection in April and May; alfo for feals, fea lions, birds eggs, &c. The 1ft of June whales and other fifh are caught, farana gathered, &c. The firft fifh that appears is the halibut; then falmon, the fame fpecies as in Kamtfhatka. They continue this chace till the end of October, when they retire to their winter dwellings. November they fpend in vifiting each other, feafting in the manner of the Oonalafhkans, and dancing with mafks and painted faces.

A VOCA-

A Vocabulary of their Language, as well as of the languages of the other nations that I have vifited, is given at the end of the Volume.

The birds that I obferved hereabout were fuch as I faw at Oonalafhka, and about Shumagin's iflands : wild geefe *; different kinds of gulls ; the crefted and tufted auk ; blue pettrel, of a rufty dark brown, very like the fwallow ; the foolifh and black guille-mot ; divers, and a great variety of ducks : the flefh of which are eaten by the natives, the fkins ufed for dreffes, and the bills, particularly of the fea-parrot, employed for ornament.

Bears now and then appear upon the ifland of Kadiak, fwim-ming acrofs the ftraits that divide it from Alakfa five miles. The whiftling marmots are numerous, as are alfo mice. Foxes, and ftone-foxes, are fcarce fince the eftablifhment of the Ruffians ; in fact, thefe and the marmot are the only animals that the Ruffians can kill ; for they are not capable of chafing the fea animals, which requires particular agility in governing the fmall leather canoes, in which the natives purfue the fea-lion, the urfine-feal, fea-otter, porpoifes, and common feals.

The fea-lion, called by the Ruffians fivootfha, is the ftrongeft and largeft of the feal kind ; covered with dark coloured coarfe hair, which is very thick and long about the neck and fhoulders ; the hind part is tapering, with fmooth fhort hair. The largeft is about eight feet long. They copulate and pafs every night on

---

* Goofe with a black bill ; the upper mandible has a callous elevation. A triangular white fpot runs from the throat along the cheeks on both fides, to the hind part of the head. The bottom of the under part of the neck, vent feathers, belly, and coverts of the tail, white ; breaft, back, and wings dufky brown ; legs a dull dark colour.

fome

some rock by themselves, one male and a number of females, driving away, or killing, every other species of animal that may approach them. The males have frequently very desperate engagements, and the conqueror is immediately joined by all the females. They are extremely bold, and will attack men if disturbed on the rocks. They have a small white spot on the temples, nearly as large as a half-crown piece; and this is the only place about them vulnerable by arrows, which hardly pierce the skin in other parts; but, if poisoned, they penetrate deep enough to infuse the baneful quality. The meat of these animals is cut in thin shreds, and dried by the hunters, who esteem it good eating. I thought it bad and fishy; but the head, which is equal in size to that of a large ox, I thought very good, if well stewed, and eaten with sarana and other edible roots. The second species is the kotic, or ursine-seal: the largest are about six feet long, covered with beautiful silvery grey hairs, of the colour of the Siberian squirrel, having a soft downy under fur, resembling brown silk. The young kotic are extremely playful in the water; the head very nearly resembles that of a lamb with long ears; and they live upon rock-weeds. The flesh of the young ones is well tasted, but the colour is blue, and unpleasant to the eye. These swarm together in great herds on the low islands, and are killed by being struck just above the nose with a short bludgeon. When they find themselves in danger, they attempt to bite. When very young, the fur is of a beautiful short glossy black, which changes to silvery when they grow up; and when they become very old, they are almost white.

The most valuable fur is that of the sea otter, called by the hunters, and in Russia, Morskoi Bóbre. The fur of the young ones is rough and long, of a light brown colour (something like the young cub of a bear), and is called Medvedka, the diminutive

of

of bear: this is of no value: the middling fized are darker and valuable; thefe are diftinguifhed by the name of Kofhlok: but the moft valuable are what is called the Matka, or mother; the largeft are about five feet long, with a rich fur nearly black, interfperfed with longer hairs of a gloffy white. The fur is upright, not inclining any particular way, from an inch to an inch and a half long. I had a young fea otter dreffed, and it tafted exactly like a fucking pig. There are no more on the coaft of Kamtfhatka; they are very feldom feen on the Aleutan iflands; of late, they have forfaken the Shumagins; and I am inclined to think, from the value of the fkin having caufed fuch devaftation among them, and the purfuit after them being fo keen, added to their local fituation between the latitudes of 45° and 60°, that fifteen years hence there will hardly exift any more of this fpecies.

Sea cows were very numerous about the coaft of Kamtfhatka, and the Aleutan iflands, at the time when they were firft difcovered; but the laft of this fpecies was killed in 1768 on Bering's ifland, and none have been ever feen fince.

Whales are in amazing numbers about the ftraits of the iflands, and in the vicinity of Kadiak; the natives purfue them in their fmall boats, and kill numbers with a poifoned flate-pointed lance. Their melted fat is an article of great trade to the continent, being carried thither in bladders by the iflanders; for which they obtain the land animals, boats, darts, flaves, &c.

I obferved the fame fpecies of falmon here as at Ochotfk, and faw crabs; fome fhells of lobfters in the beach; cockles weighing a pound each, and a variety of other fhell-fifh. Thefe are the food of the fea-otter.

The

The Halibuts in thefe feas are extremely large, fome weighing feventeen poods, or fix hundred and twelve pounds avoirdupois. The fins and tail are good eating; but the body of the fifh is very coarfe and dry. The liver of this fifh, as alfo of cod, the natives efteem unhealthy, and never eat, but extract the oil from them.

The harbour in which the Ruffians have their Eftablifhment is called Treeh Svatiteley. It is on the fouth-weft fide of the Bay formed by a low fpot of land running out from the fide of one of the loftieft mountains; and, taking a circular fweep north and weft, forms a harbour of about two miles in circumference, with foundings from eight to three fathom, over a bottom of mud. Near the dwellings, is a frefh water brook iffuing out of the mountain; and at the bottom of it are their cook-houfes, and two infignificant falt water lakes.

This ifland is fubject to frequent earthquakes, which are fometimes very violent. We obferved high water at the new moon at 11° 45', the rife about eight feet. The variation of the compafs 26° eaft by the meridian line: the longitude of the harbour 205° 30', latitude 57° 5'.

The natural productions of the ifland that fell under my view were, the elder in abundance; the low willow; fome brufh-wood, ginfeng, wild onions; the edible roots of Kamtfhatka; feveral fpecies of berries, with currants and rafpberries in abund-ance, the latter white, but extremely large, being bigger than any mulberry that I have ever feen, but watery in tafte. Several of the natives had fmall bunches of fnake-root, which they obtained from Alakfa.

In

View of Sheldkoffs Establishment on the Island of Kadiak.

1 Dwelling Church.
2 Astronomical Tent
3 Gallows build on shore.

The North extremity of the Island Tanaga bearing South distant 15 Miles.

Island of Atcha 22 June 1791 bearing S.E.

In the interior, they have good timber of common pine ; and on the eaftern point of the ifland, which Captain Cook called Cape Greville, they have a very confiderable foreft of pines, whence they bring the trees to build their huts here, and repair their veffels.

CHAP.

CHAP. XIV.

Leave Kadiak.—Island of Afognak.—Shuyuch, or Point Banks.—
Icy River.—Fall in with a groupe of Islands, and are visited by
some of the Natives.—Anchor in Prince William's Sound.—Visited
by the Natives, whose propensity to thieving is checked by the
sagacity of two Dogs.—Captain Billings assumes an addition-
al rank.—Captain Saretsheff's Account of his Survey of the
Coast.—Cape St. Elias ascertained by an Extract from Mr. Stel-
ler's Journal.—Mr. Delareff's Account of a former Visit to the
Sound.—Some Reflections of the Author.

HAVING remained here at anchor until the 6th July, we took
on board our astronomical tent, and the tent containing our travelling church, which were both erected on our arrival; and at five
o'clock in the evening of that day, we hauled out of the harbour
into the bay, and experienced as much difficulty in getting out, as
we had before done in going in.

Mr. Delareff, the director of the Company, upon receiving intelligence that a Spanish frigate under the command of Captain
Mendoza was at the entrance of Cook's river, acquainted us,
and took his passage on board our ship. We were informed,
that the Spaniards were in the habit of visiting the settlements
yearly, and that the Russians obtained some provisions, and a considerable quantity of sea-otter skins from them, in exchange for
hardware, beads, and linens.

It

It was Captain Billings's intention to visit this ship. We had variable light airs from north to west, so that we made but little headway; and observed the whole of the south-east and east shores of the island very lofty and broken, replete with inlets and bays; and numberless rocks close in with the land. The 7th, at 5° 22′ 55″ P. M. our longitude 207° 47′ 45″, latitude 57° 25′ 40″, Yellovoi Muis, or Cape Greville, bore north-west.

The 8th, at day-break, we saw the island of Afognak, upon which the Russians have an establishment. This island is covered with fine timber inland. Its distance from the north extremity of Kadiak is seven miles. The straits are replete with islands and rocks. Two miles north of Afognak is the island Shuyuch, surrounded with rocks, and about four miles in length. Its northern cape was seen by Cook, who named it Point Banks.

Contrary and baffling winds from the west and north-west prevented our weathering Cape St. Elizabeth. In the morning of the 11th we had rainy weather, but saw in the haze the land east of the cape, which was much broken, and mountainous. At 10 o'clock a conical mountain on the continent bore north-west 22°. We observed a river, which the hunters call Ledenaia Reka (Icy River), from its being continually frozen, and which serves them for a direction into the Sound. It bore north-west, distant about 15 miles, and is situated in latitude 59° 36′, longitude 209° 45′ east. Near the mouth of this river are a groupe of islands, and numerous detached rocks. We had variable light airs from the south-east to north-east, with calms, and foggy and misty weather, which hid the land till Friday the 12th at two P. M. when we again saw land about 15 miles ahead. Our course was north north-west, with light airs from the north-east. Observing se-

B b

veral

veral openings, and Delareff affuring us that there were no funken rocks, but good anchorage in the bays, we ftood in, and got among a variety of fmall iflands. When we were at about three miles from fhore, two of the natives came off in their canoes, making the general fign of peace, by expanding their arms ; we repeated their fignal, and hoifted a flag, upon which they came on board, bringing with them the fkin of a young fea-otter, a river-otter, and a feal ; for which they received tobacco and beads. They ftayed but a fhort time on board ; and Delareff went on fhore in his three-feated baidar, which he brought with him, and two Americans, accompanied by the natives who had vifited us ; he not feeing any probability of getting with the fhip to Cook's river. Though quite unarmed, he did not apprehend any danger ; a plain proof that his company had the complete friend-fhip of the natives. We defired him to acquaint the Spaniards, that we wifhed much to fee them, and fhould continue fometime in Prince William's Sound. He gave us a young American, who underftood the Ruffian language, for our interpreter. This young man fpoke with the two natives, who faid, that almoft all the land which we faw was iflands, which produce very fine timber to the water's edge. Our latitude was now 59° 15'; and, though we were not above three miles from the neareft land, quite em-bayed, we got no foundings with 100 fathom line. We ob-ferved the current fetting to the weftward at two knots and a half. Variable light airs between fouth and eaft continued, with calms all day. The 13th, at 4° 4' 25" P. M., our longitude was 209° 15', latitude 59° 17' 45". Baffling winds continued till the morning of the 16th, when a moderate breeze fprung up from the fouth eaft. Our courfe was north-eaft, at about five knots, with a heavy fea and rainy weather. In the afternoon we had again little wind. In the morning of the 17th, at four o'clock,

we

we faw Montague ifland (called by the natives Tfukli) bearing north 5° eaft, diftant about feven leagues. At 8° 36' 25" A. M. our longitude was 211° 13', latitude 59° 43' 38". We ftood away to the eaftward of the ifland, to get into Prince William's Sound, which we accomplifhed on the 19th at four P. M., and brought up near the place where Captain Cook lay at anchor in 1778.

In the morning of the 20th July, we fent our obfervatory, with all the apparatus, on fhore, and were vifited by numbers of the natives, who were at firft very fhy; rowed about the fhip holding up their hands with bear-fkin gloves on; finging, and making figns of friendfhip. On being affured of friendly treatment, they foon came on board, and manifefted a ftrong inclination to fteal iron articles. They complained bitterly of the ill treatment that they had received from a Ruffian veffel under the command of Sturman Polutoff. Thefe Ruffians had taken their fea-otter and other fkins from them without making any returns; wantonly fhot fome of their people, and carried feveral of their women away by force.

They feemed perfectly fatisfied with the treatment and fome prefents which they received on board, and left us with a promife of returning with fome fkins. However, they made fhift to take with them every thing that lay about carelefsly, and the iron tiller of the boat along-fide.

We kept a ftrong guard on board, and alfo at the obfervatory, to prevent our being furprifed; for the natives appeared refolute enough to undertake any thing.

B b 2

A num-

A number of them vifited Captain Billings in his tent on fhore, and he treated them with tea; of which, however, they did not feem very fond, nor of brandy, nor tobacco; for thefe articles had not yet been introduced among them, or tafted, except by a very few who had been on board the Ruffian veffels; but they were all fond of fugar.

A water fpaniel that Captain Billings had with him did not feem to like the appearance of thefe favages; however, he lay ftill in the middle of the tent. The cabin-boy had carelefsly placed the tea-board fo, that part of it, with fpoons, &c. was feen on the outfide of the tent. One of the natives attempted to appropriate the fpoons to himfelf; this no one obferved but the dog, who fprang up, leaped over the natives in the tent, feized the thief by the hand with the fpoons in it, and held him faft till the Captain told him to let go: a circumftance which, I believe, kept them honeft afterwards in the dog's prefence. Captain Hall had a pointer on board, which did the fame fervice there. The natives wifhed very much to poffefs thefe dogs, and one of them defired Captain Hall to fell him half of his; which induced me to think that they wanted to eat them, or fuppofed that they were kept to be eaten by us.

Captain Billings, being now arrived at the place which he fuppofed to be Cape Saint Elias, difcovered by Captain Bering in 1741, affumed an additional rank, conformable to the mandate of Her Imperial Majefty, and took the oath adminiftered by our prieft, according to the rules of the fervice. This was upon Saint Elias's day.

Two men and a woman had accompanied us from the island of Oonalaſhka, by their own deſire, to ſerve as interpreters. Their chief view was, to get out of the way of the Ruſſian hunters now on their iſland; and Captain Billings promiſed to leave them at home on his return, when they thought the hunters would be gone. They had brought their ſmall canoes, or baidars, with them. I was the only perſon on board, except the Aleutes, that could venture out in theſe boats; and the 22d, being a fine day, with light airs and calms, I took a ſmall excurſion merely for exercise, quite alone; but received Captain Hall's injunctions (Captain Billings being at the obſervatory) not to go on ſhore, nor venture to any great diſtance. I left the ſhip at one o'clock, and paddled with the tide at the rate of about eight miles in the hour, without paying any attention to the diſtance. On attempting to return, I found the tide too ſtrong againſt me. I did not ſee a ſingle native any where, nor any traces of them, and reſolved to enter a ſmall cove to wait the return of tide, and to get a draught of freſh water from a brook that I obſerved. After entering a ſmall inlet, I diſcovered that my retreat was cut off by ſome of the natives. My dreſs was a nankeen jacket and trowſers; and I had a few claſp knives and beads in my pocket, which I gave the natives; particularly a woman whom I obſerved amongſt them in a nankeen camley, and who addreſſed me, to my aſtoniſhment, in the Ruſſian language; which rather increaſed the uneaſy ſituation that I found myſelf in, on account of the complaints that they had made, on board, of Polutoff's company. I found, however, no great difficulty in perſuading her that I was not a Ruſſian. She gave me a bowl of water, and treated me with berries upon which the oil of ſeals had been poured. She told me, that Polutoff had taken her away by force, and kept her above a year, till ſhe had learned the Ruſſian language. After

that,

that, she associated with Zaikoff, and returned to the Sound, making herself their interpreter. She said, that Zaikoff, who was a very good man, and behaved well to every body, had favoured her escape, and that they had been well revenged upon Polutoff and his crew; for that a boat from each of the vessels had been on shore to cut wood, and had pitched two tents (one for each company) at a small distance from each other. It was in the autumn; the night was dark; and only one man watched at a fire side, sitting on the beach. The natives crawled, unheard, close to the watch at Polutoff's tent, killed him, and, rushing into the tent, murdered every soul there, without molesting Zaikoff's tent, or any of his people.

She invited me to their dwelling, and assured me that I should be safe. I asked her how far it was. She said, that if I left the ship at sun-rise I should arrive at her dwelling before sun-set; that the habitation was across the straits at the end of the Sound (pointing to the eastward of the north), near the discharge of a large river. This induced me to ask her, if the land about us constituted any part of the continent. After some conversation between her and the chief, she told me, that the openings were all straits. I promised that I would go with her if they would come on board in the morning for me, and that I would give them beads and other trinkets. At half past three it was high water, and I put off, very well pleased to get away; for they all admired my baidar so much, that I was much afraid of losing it, and my sensations, when I first discovered myself in their power, were very unpleasant. I arrived on board at half past four, and relieved Captain Hall of his anxiety on my account, but forbore relating my adventure, lest it should prevent my future excur-

sions,

fions, which I promifed myfelf fhould not lead me into fuch danger a fecond time.

Early in the morning of the 23d the woman came alongfide, with about ten double canoes, and brought a fea-otter fkin, which I took for a few beads. They afked me to accompany them, and the chief would remain in the fhip till I came back; but Captain Billings would not agree to it. Neither Captain Hall nor Saretfheff faw any reafon for objecting to this trip, efpecially as the chief offered to ftay on board as hoftage for my return. Captain Billings at this time had the woman and chief in his cabin, out of which they returned in great hafte, and in feeming rage left the fhip. I was extremely forry, as it deprived me of the hopes of getting fuch information as I wifhed to obtain concerning the ftraits, and particularly the large river that fhe fpoke of. They rowed to the obfervatory, and took a cafque from the head of one of our grenadiers, with which they attempted to run away, but returned it on being overtaken.

They, indeed, fhewed an aftonifhing propenfity to thieving, even of fuch things as could not have been of any fervice to them; and, upon being detected, returned the articles with amazing compofure. Their language and manners differ but very little from thofe of the iflanders at Kadiak.

Towards evening of the 24th July, Captain Saretfheff went with the long-boat armed, to furvey the Sound, to examine the dwellings of the natives, and to difcover whether the land was any part of the continent, or merely iflands. He returned in the afternoon of the 27th, and gave the following account of his excurfion.

" I went

" I went north about eleven miles and a half, where the coast
" trends eastward. Here six Americans, in four baidars, over-
" took us, and said that they wanted to conduct or accompany
" us. In the evening, when we halted for the night in a small
" bight (not so far as they wished us to go), they left us. Here
" we saw a cross affixed *. The next day we proceeded; and
" at the distance of sixteen miles and a half the land trended
" away to the south-east, into what appeared to us a very exten-
" sive bay. On account of foggy weather, we could not well
" discern the opposite shore, which, however, appeared at times,
" and seemed rather low land. At eighteen miles the shores led
" to the north-east. Here we again saw the natives in eight
" baidars. They said that they had been on the chase out at sea,
" and that we were in the straits; but advised us not to continue
" our course much farther, as it was very shallow, and the
" breakers were so violent that they found great difficulty in
" passing in their small canoes; adding, that the place was quite
" dry at low water. They said, that the opposite shore was like-
" wise a large island, and that the straits were also shallow and
" nearly dry at low water. The opposite shore was not to be
" seen on account of the fog. We proceeded, in all, twenty-
" three miles. The fog clearing up a little, discovered both
" shores and the sea. The cape on the right hand was about
" two miles distant. The left shore trended to the north-east.
" At a little distance from shore were two small islands, and a de-
" tached high rock. We crossed the straits backwards and for-
" wards in returning, with soundings from one and a half to two
" and a half fathoms, sand. It was high water; and, that the

---

* I am inclined to think that it was erected by Zaikoff, or Polutoff, in consequence of
their people being buried here.

" boat

" boat fhould not be left dry, we proceeded back at feven o'clock
" in the evening, and paffed the night of the 26th in a fmall
" bight about fix miles from the fhip. Here we found a few na-
" tives in their fummer habitations for the fake of the chafe,
" who received us in a very friendly manner, affuring us that
" they had no bad intentions, becaufe we behaved well to them,
" and not like fome vifitors who had been before us. I told
" them, that thofe who treated them ill were not government
" fhips; and that whenever they faw a fhip with fuch a flag as
" ours, they might go on board with great fafety." (* Signed
G. Saretfheff, and dated 27th July 1790.)

He did not like the appearance of fome of the natives, and
kept a very good look out, to prevent his being furprifed.

On the 28th, I. made a little excurfion in the long-boat well
armed, with the naturalift and drawing-mafter, and returned the
next day without feeing any of the natives, or meeting with any
circumftance worth relating. An old man came aboard on the
29th, who feemed very good natured and intelligent. Mr. Saretfheff and I entered into converfation with him through our American interpreter, and afked him, how long it was fince the firft
fhips made their appearance among them; and whether he remembered any boats having been loft? He anfwered, that feveral
boats had been loft, which, by his account, we thought to have
been Spaniards. He faid, that they frequented (on the chafe in

* I think it neceffary to notice, that upon Mr. Saretfheff's arrival in Kamtfhatka he difcovered that his interpreter knew that the natives wanted him to go up the bay, that his
boat might be left dry, when they meant to attack him and murder all his people. Upon
Mr. Saretfheff's afking this interpreter why he did not mention it at the time, he faid, " I
" fhould have been fafe had you been murdered; but, had I difcovered their plan, I fhould
" certainly have been killed."

C c

fummer) an ifland, which he defcribed fo particularly, as convinced us beyond a doubt, that it was the Kay's ifland of Captain Cook. He remembered, that when he was a boy, a fhip had been clofe into the bay on the weft fide of the ifland, and had fent a boat on fhore; but on its approaching land the natives all ran away. When the fhip failed, they returned to their hut, and found in their fubterraneous ftore-room, fome glafs beads, leaves (tobacco), an iron kettle, and fomething elfe. This perfectly anfwers to Steller's * account of the Cape Saint Elias of Bering, and

* The following is a tranflation of this part of Mr. Steller's journal, which he kept in the German language.

" We faw land the 15th July; but, as it did not appear diftinct enough to make a drawing of it, it was, on account of my having feen it firft, faid to be a miftake; but the next day it appeared beyond a doubt. The land was high, and an interior mountain was very plainly difcernible fixteen German miles out at fea. I have never feen, in all Siberia or Kamtfhatka, a more lofty mountain. The fhore was broken every where, and difcovered numbers of inlets and harbours. Every perfon congratulated the commander on the difcovery; which congratulations he received not only with aftonifhing indifference, but even fhrugged up his fhoulders, and faid to Mr. Plinifner, " We imagine that we have found " every thing, and numbers are grown big with airy projects. Nobody confiders *where* we " have found land, the diftance that we have to run back, or what may happen: per- " haps paffage-winds may prevent our return. We know not this land, nor have we fuffi- " cient provifion to pafs a winter."

" The 17th, on account of little wind, we advanced flowly. The 18th, towards evening, we came fo clofe as enabled us to fee plainly the beautiful forefts that approached to the water's edge. The fhore was even, and appeared fandy. We kept the continent on the right hand, failing a north-wefterly courfe to get behind an ifland confifting of a high mountain covered with wood, which was only to be done by plying to windward.

" The 19th we were two German miles off the north-weft extremity of the ifland. We had obferved, the day before, ftraits between it and the continent; and I thought that fome confiderable river emptied itfelf in the vicinity. My reafons were,—the current two miles out at fea; and the difference in the colour of the water, which was alfo frefher.

" I mentioned my conjectures, but they were laughed at. The whole of this day we employed in plying to windward, to get clofe to the ifland into the inlet that we had obferved the

and is undoubtedly the very fpot where Steller landed, and where the things above mentioned were left in the cellar. Thus it is very plain, that Cape Saint Elias is not the fouthern point of Montague ifland, but Kay's ifland. This native farther told us, that at the north extremity of Kay's ifland, there was a bay fheltered from the wind; that the entrance at low water was as deep as his double paddle (which is about feven foot); and that there are runs of frefh water into it, but no great rivers. A very confiderable river, however, falls into the fea a day's journey north of our anchorage, up which the natives travel 14 days to the refidence of a different nation, the people of which fupply them with knives, copper kettles, and inftruments, and make their canoes.

the day before. The 20th we came to anchor between iflands; and, in compliment to the day, named the extremity of the large ifland Cape Saint Elias," &c.

Chytroff, the mafter, was fent on fhore to furvey, and Mr. Steller accompanied him to make his obfervations on fhore on the " three kingdoms of nature."

He faw the traces of inhabitants, and difcovered one of their cellars, into which he entered. It contained,

1. Lukofhkan. Thefe are a kind of box of the bark of trees, about two yards high, containing fmoke-dried falmon.

2. A quantity of the fweet plant of Kamtfhatka, but cleaned and prepared in a better manner.

3. Several fpecies of grafs, cleaned like hemp: I took them for nettles, which grow here in abundance; perhaps ufed, as in Kamtfhatka, for fifhing-nets.

4. The dried inner bark of larch and fir in rolls, fuch as I have feen in Kamtfhatka, through all Siberia, and even in fome parts of Ruffia; and which is eaten in cafe of need.

5. Large packs of thongs of fea-weeds, of great ftrength.

Befides a few arrows made like thofe of the Tartars and Tungoofe; blacked, and wrought fo fmooth, that I apprehend they have iron inftruments.

He carried with him on board two bundles of fifh; an arrow; a wooden inftrument for making fire, refembling that ufed formerly by the Kamtfhadals, with tinder made of dried leaves; a bundle of the wood; fome bark, and fome of the grafs.

Sailors were afterwards fent to leave an iron kettle; a pound of tobacco; a Chinefe tobacco pipe, and a piece of Chinefe filk; in return, they nearly plundered the cellar.

That

That thefe people trade with others farther inland, and obtain from them knives and other articles; but that his nation never go farther than 14 days' journey. That the articles of their trade are, the fkins of fea-lions, for boats; oil of fea animals; fmall fhells; and mufcle-fhells for points to arrows; and that thefe were a very powerful and warlike people.

Another obfervation of his, I think it very neceffary to mention: it was a pofitive affertion, that there were ftraits and iflands as far as we could fee; and that to the fouth-eaft there was " A GREAT SALT WATER," with many entrances to it. I repeatedly afked the queftion, and could not be miftaken in the anfwer; and I would moft willingly have ftayed on the coaft alone, to explore thefe unknown parts from tribe to tribe, until I had loft myfelf, or found my way to Europe through fome of thefe cranny paffages. I am aware, that I was thought a madman for it; but this madnefs, this enthufiaftic confidence, would, I am certain, have affifted my fuccefs; nor would I have left unexplored a river of which we had fuch confirmed accounts, without good reafon for it; for I never met with any men that would refufe affiftance to one individual, who, without the means of being their enemy, was at all times in their power. Over and above all this, I declare, that I have complete confidence in a Supreme Being, who governs every thought, and infpires means of expreffion to fecure the devotee in exploring his wifdom.

I hope that my rhapfodies will not offend my readers: they are notes penned at the inftant when my feelings were moft acute, and with a view of making them known to the public on a future day.

Captain

Captain Billings had received intelligence of this river from Mr. Delareff, the director of Shelikoff's companies at Kadiak, Afognak, and Cook's River; who gave the natives the character of good people; and faid, that they ate, drank, and flept together in the moft friendly manner; and I firmly believe what he faid *.

We took in a number of fine fpars, with a fupply of water, and caught with our net in-fhore fome falmon, befide taking fome flat fifh by angling over the fhip's fide. Having hauled up a large fkate while the natives were about the fhip in their boats, as many

---

* I think it neceffary to communicate the following intelligence of this Gentleman *verbatim*.

" I failed from Ochotfk in the month of July 1781; arrived the 10th Auguft at Commandorfki (Bering's) Ifland, where I wintered. The fecond winter I paffed at Oonalafhka; and the winter of 1783 at Prince William's Sound. I arrived on the 13th Auguft in the offings before fun-rife, and fent out a boat well manned to feek a convenient harbour. The weather became thick and hazy; but when it cleared up I difcovered a number of boats making to the veffel. The largeft among them hoifted a flag; I did the fame; they then rowed three times round the veffel, one man ftanding in the middle, finging, and waving his hands. Upon being invited they came aboard, and I obtained fourteen fea-otter fkins for glafs beads, chiefly blue. I offered them fhirts and clothes, which they did not feem to want; and tobacco they rejected. They behaved in a very friendly manner. We were quite off our guard, and ate, drank, and flept together in the greateft harmony. They informed us, that two fhips had been there fome years back, and gave them great ftore of beads and other articles. By their defcription, thefe veffels muft have been Englifh. They had knives and copper kettles, which they faid they obtained up a great river, about 14 days journey againft the ftream, where the natives were numerous, and had great quantities of copper.

" On the 8th September there feemed fome alteration in their appearance, and they fuddenly commenced a violent attack upon my people. I knew no caufe, until the 21ft, when the boat fent upon my arrival returned, and I found that there had been quarrelling and fighting between the boat's crew and the natives. I could never find out the origin of the difpute; but really think that my own people had been the aggreffors; perhaps for the fake of a few fkins. Polutoff's veffel was at that time in the Sound, and I left them there."

I   as

as could get at it ftabbed it with their fpears with great eagernefs, and called it the devil.

Rafpberries were in great plenty, white, extremely large, and fine flavoured. Cranberries and feveral other fpecies of berries we obferved, with plenty of ginfeng, and fome fnake-root. The timber comprifed a variety of pines of an immenfe thicknefs and height; fome extremely tough and fibrous, and of thefe we made our beft oars.

The natives wore the fame habits as thofe of Kadiak; they poffefs the fame cuftoms, and the languages differ very little. They had evidently a knowledge of feveral European words; for if they were not fatisfied with returns made in barter for their articles, they exclaimed, *No! no! no! no!* holding their hands for more; and if more was required from them than they inclined to part with, they fhook their heads and faid, *Plenty, plenty.* They obferved the expreffion of our countenances very minutely; and if they faw any thing refembling anger, they immediately laid down their articles in their boats, held up their hands, and exclaimed, Amigo, Amigo! and La-lee, La-lee! which they underftood to fignify friendfhip and peace. Their inftruments differ only in this particular, that many of them are pointed with copper, and one of them had an European bayonet on the end of his fpear. They have very large fcreens; I was told, (but faw none) of fufficient ftrength and thicknefs to withftand a mufket-ball, and large enough to fhelter twenty or thirty men. They have armour of wood, which covers the body of the warrior and his neck; but his arms and legs are expofed. This is made of very neat pieces of wood, about half an inch thick, and near an

inch

inch broad, tied very artfully together with fine threads of the finews of animals; and fo contrived, that they can roll it up or expand it. This they tie round the body, a flap before reaching down their thighs; but fo made as to rife or fall, and permit their fitting in baidars: a fimilar flap hangs on the breaft, which may be rifen as high as their eyes. Straps faften this armour on their fhoulders, and ftrings tie it round the body on one fide.

The head is well guarded with a wooden helmet; fome of thefe are made to refemble the head of a bear, and cover the face completely. Such wooden caps, or head-pieces, are worn in the chafe of the different animals which they reprefent; the native clothes himfelf in their fkins, and approaches within a convenient diftance to ufe his bow or lance. Some of the natives were prefented with copper medals and beads.

Our obfervations proved the longitude 213° 42′ 45″, latitude 60° 18′ 48″, at the obfervatory on fhore, about 50 fathom fouth of the fhip at anchor.

CHAP.

CHAP. XV.

Leave Prince William's Sound.—It is refolved to return to Kamt-
fchatka.—Kay's Ifland.—One of the Aleutes taken on board from
Oonalafhka makes an attempt on his life.—Extraordinary diffe-
rence in Longitude between the Time-keeper and the Ship's rec-
koning.—Short Allowance enforced.—Arrive in the Harbour of
St. Peter and St. Paul.

WE remained in this ftation till the 30th July 1790, at fix
A. M. when we fet fail, having hauled out of the bay with a
gentle north-weft wind, fhaping our courfe fouth and fouth-eaft.
At noon the north extremity of Montague ifland (called Tfukli
by the natives) bore by true compafs north-weft 40°, diftance
feven miles. From this point, latitude 60° 16′, longitude 213° 3′,
we took our departure. Variation of the compafs 28° 30′ eaft.

In the morning of the 31ft July we faw Kay's ifland, and the
detached rock off its fouth extremity. At 1° 26′ 47′ double alti-
tudes made the latitude 59° 51′ 22″, when Kay's ifland bore
north-eaft, diftance about eight miles. At four we faw plainly
Mount Saint Elias bearing north-eaft 49°. At five P. M., longitude, by time-keeper, 215° 42′ 45″, latitude 59° 44′ 22″, variation of the compafs 26° eaft. The weather foon after grew hazy,
and we faw no more of the land. The wind fhifting to the eaft
and north-eaft, we kept all the 1ft of Auguft a fouthern courfe;
taking into confideration our fmall ftock of provifions, which precluded

cluded every thought of paffing the winter where we could not be fure of procuring a fupply; together with the latenefs of the feafon, and the diftance that we had to run back to Kamtfhatka; befides, it was thought neceffary to have a fecond veffel, for fecurity's fake, in fo uncertain a navigation, where none of the iflands, except Oonalafhka, were laid down with any fort of exactnefs even on the beft of charts. Our return, therefore, to Kamtfhatka was, neceffary to forward the bufinefs of building the veffel. It was alfo confidered as a principal object of the expedition, to obtain fome more perfect information concerning Cook's River, and other rivers and parts of the continent fouth of it, as well as to furvey all the chain of iflands between America and Kamtfhatka, and afcertain by aftronomical obfervations their true fituation. To effect which, the whole of the next fummer and winter might be employed, and the fummer following appropriated to explore the more northern parts to the utmoft extent of poffibility.

This was agreed to, and our return determined upon immediately; intending, however, to ftop at Oonalafhka for water, and to put on fhore the natives who had accompanied us.

I believe that I was the only perfon on board who felt any regret at the thoughts of returning to Kamtfhatka. I really imagined that we fhould never fee this coaft again; and I had now acquired knowledge enough to furvey any place, from Mr. Saretfheff's intelligent manner of explaining whatever appeared to me difficult. This, added to a few leffons that I took from Mr. Batakoff, our mafter, would have enabled me to be pretty exact; a confideration which made me offer to go on fhore alone, and meet Captain Billings the enfuing fummer at any part of the coaft that he would appoint. The attempt might have been rafh: I do not, however, think (as I before obferved) that one perfon runs any rifk either of ftarving, or being murdered, but may depend upon fure conveyance from one tribe to another. I do not mean to infer that there was any impropriety in rejecting my offer, for it was regarded as facrificing myfelf to no purpofe.

Kay's ifland, the laft that we faw to diftinguifh as fuch, (the fouthern point of which moft affuredly forms the Cape Saint Elias of Commodore Bering,) is very remarkable. It is of moderate height, except the fouth extremity, which is confiderably elevated above the reft, and terminates very abruptly a barren mountain of a faddle-form and white. A detached rock of the fame kind of ftone is fituated a few fathoms off the point; eaft of which, at the diftance of one mile and a half, are funken rocks. The other part of the ifland confifts of hills and vallies, apparently well wooded with fine pines. From the fouthern point, the ifland trends north 46° eaft, twelve miles in a ftrait line, and is two miles and a half acrofs in the wideft place. To the weft of the northern extremity is another ifland, with feveral fmaller ifles nearer the continent; forming a well fheltered bay over a bar of about feven feet at low water, with a rivulet at its head. The direction of Mount Saint Elias from Kay's ifland is eaft north-eaft. It towers to an immenfe height, and is covered with fnow. Its diftance I compute at about 30 leagues.

If I may be allowed to hazard a conjecture of my own con-cerning the land that we faw, it is, that I do not think any one place, except Mount Saint Elias, conftitutes any part of the conti-nent; not even Cape Elizabeth; and I have my doubts of Alakfa itfelf. I think that the whole is formed of a clofe connected chain of iflands, feparated by ftraits from the main land. I ob-

I

ferved

ferved no change in the colour of the water, however clofe in with fhore; which muft have been the cafe had any confiderable rivers fallen into it; but we faw none, and our enquiries do not juftify the fuppofition that rivers exift, except beyond the ftraits; for the rivers were fpoken of by the natives as lying behind the iflands. I could not perceive any alteration in the tafte of the water, not even where we were at anchor, and it was exceedingly pellucid.

However, I fhall take leave of this coaft, and proceed to give an account of our return. No fooner was this refolved upon, than the wind fhifted from north-eaft to weft and fouth-weft. We kept a courfe as much to the weft of the fouth as poffible; and on the 4th, at eight P. M. our time-keeper gave the longitude 215°, latitude 56° 53′, variation 27° 50′. On the 6th, we faw feveral land birds and floating wood, our latitude 55° 15′, longitude 214° 15′, variation 26° 10′ eaft. We had a brifk gale at fouth-weft, fteering fouth fouth-weft, one half weft, the fea running very high, fucceeded by calms and variable light airs: all the 9th we had a favourable breeze from the fouth fouth-eaft, with rainy and mifty weather. We made a good run weft fouth-weft. On the 10th, calms and baffling winds. At 4° 10′ 25″ P. M. apparent time, our longitude by time-keeper was 210° 9′ 15″, latitude 54° 29′ 17″. The fhip's reckoning made us one degree more weft; but the latitude was within two miles. The difference of longitude increafing every obfervation induced Captain Billings to doubt the rate of going of the time-piece. On our paffage out, the fhip's reckoning and our obfervations agreed fo well, that he could not by any other means account for the difference. On the 14th, the amplitude of the fetting fun gave the variation 23° 12′ eaft. The 15th, at nine A. M. longitude

by

by time-keeper 201° 49' 30", latitude 54° 15' 6", by ship's reckoning longitude 200° 47', latitude 54° 11'. Misty weather prevented our observing again till the 26th August, at 9° 10' 45" A. M. when the time-keeper gave the longitude 194° 21' 15", ship's reckoning 190° 20' east. At noon, our observed latitude was 52° 22' 16", by ship's reckoning 52° 14'. In the evening of the 26th, we encountered a hard gale of wind from west south-west, and laid-to under fore, main, and mizen sails. The gale continued till the 28th, at five P. M. when, it abating a little, we set our close-reefed top-sails. At noon we got an imperfect view of the sun, which gave the latitude 52° 13' 2", variable light airs, cloudy: at times rain. Supposing ourselves about the meridian of Oonalashka, having but a scanty supply of water on board, and the natives wishing to be at home, we stood to the north for this island, with rainy and foggy weather, which grew so thick that we could not see half a mile a-head; and reaching the latitude of 52° 59' by account, without seeing land, at the same time a brisk gale springing up from the south by east, it was resolved upon to prosecute our voyage to Kamtshatka, and keep the natives till the next spring. The use of fresh water was now prohibited, except where absolutely necessary: foggy, misty, and rainy weather continued. The 30th, at six P. M. we thought we saw land west north-west, but the fog hid it before we could possibly ascertain whether it was so or not: however, we stood away to south all night. Our latitude was by account 53°, longitude 191° 25'.

On Sunday, the 1st of September, we had a brisk gale from the west north-west, steering south-west. During the night, and particularly this morning at eight o'clock, the sea running very high, and the ship in great motion, we experienced a violent shaking and trembling of the vessel, as if her keel were rubbing

against

againſt an uneven bottom; it laſted ſeveral ſeconds, and we ſuppoſed it to have been cauſed by an earthquake. At noon the altitude 41° 2′ 30″, proved our latitude 52° 59′ 46″, which, by our ſhip's reckoning, was only 52° 23′. The 3d, at 8° 35′ 25″ A. M. our time-keeper gave the longitude 195° 10′, latitude 51° 10′ 33″; our reckoning, longitude 189° 50′, latitude 51° 29′. At noon, our obſerved latitude was 51° 9′ 33″.

Notwithſtanding Captain Billings doubted very much the regularity of his time-keeper, he never miſſed an opportunity of aſcertaining the longitude with it; yet he placed the greater confidence in the ſhip's reckoning. I was quite of a contrary opinion, confirmed by the difference exiſting in the reckoning of Commodore Bering in 1741, which was near 12 degrees ahead of Bering's iſland at the time when the ſhip was wrecked upon it. The 4th September, at noon, a ſudden ſquall carried away our fore-maſt a little below the cap; the top-maſt, in falling, alſo broke. The wind was ſo heavy, as to bring our ſhip's lee-gunwale under water; but on the maſt's breaking ſhe righted; a hard gale from the weſt immediately followed, and we brought-to under mizen and main-ſail till ſeven o'clock the next morning. At 8° 44′ 15″, our longitude by time-keeper was 192° 44′, latitude 50° 36′ 7″, variation 17° 35′ eaſt, making a difference of the ſhip's reckoned longitude of five degrees. She being ſo much a-head, which, of courſe, induced us to keep a ſharp look-out, and uſe the greateſt caution, contrary winds continuing, we could make but little way by plying to windward. The 10th September, in latitude 49° 9′, and longitude by time-keeper 186° 40′, we ſaw great flocks of birds flying to the ſouth.

The

The 14th and 15th we had hard gales of wind weſt and weſt north-weſt, and rainy weather; the 16th in the morning we got a ſight of the ſun; our time-keeper gave the longitude 181° 24′ 30″, latitude 49° 1′ 48″, making a difference in our ſhip's reckoning of ſeven degrees in longitude, and ten miles in latitude. During the 24 hours we ſaw ſeveral indications of land being near; as weeds, birds, &c. The 17th, calms and baffling winds, with miſty weather: a freſh eaſterly breeze ſucceeded for about 20 hours; our longitude in the morning of the 18th was, by time-keeper, 179° 22′, latitude 48° 30′; the Bay of Avatſha bearing north 65° 35′ weſt, diſtance 655 miles. We were now without bread, and had but very little water; ſo that we ſhortened the ration of the latter, and gave a ſufficient allowance of peaſe and butter, all hands voluntarily rejecting ſalt meat. A hard gale from the weſt brought us to under mizen and main-ſail for 24 hours.

The 21ſt September, at nine A. M., we obtained ſome diſtances of the ſun and moon, which gave the longitude 178° 46′ 45″. At noon our latitude obſerved was 49° 12′ 35″, which differed nearly eight degrees from our ſhip's reckoning: however, both were kept in the Captain's journal, and our ſituation was very uncertain. The 23d, one of the Aleutes taken on board at Oonalaſhka cut his throat; but not ſo effectually as to cauſe his immediate death: his companions ſaid, it was owing to his extreme grief on hearing that he muſt go to Kamtſhatka. Hard gales of wind continued from the weſt, with hazy and miſty weather, till the 24th in the morning, when we had calms and variable light airs. At ſix this morning we ſaw land bearing north and weſt, and a conical mountain to the north-weſt, diſtant about 15 leagues, which I ſuppoſed to be the eaſtern

point

point of Amtſhitka.   At eight A. M. our longitude was, by time-keeper, 180° 44′ 45, latitude 50° 50′ 10″.   The 25th we ſaw an iſland, and the ſame iſland the 26th; for ſcant winds prevented our making much head-way.   The land that we ſaw on the 24th May, on our outward paſſage, was ſo ſtrongly impreſſed on my mind, that I had no doubt of its being the very ſame that we now ſaw; namely, the iſland of Amtſhitka above mentioned.   At that time our longitude, by dead reckoning, was 179° 00′, and our latitude 51° 18′.   Our obſerved longitude, by time-keeper, was, the 25th September, 179° 11′ 45″, latitude 50° 49′; but the haze made the land appear much nearer than it was; and the fog, hiding it, prevented our aſcertaining its diſtance by correſpondent bearings.

The ſhip's reckoning ſtill differing ſo materially from that of the time-keeper, induced Captain Billings to reject this method of aſcertaining the longitude; but he continued occaſionally to take the ſun's altitude, without making any minutes in the journal; doubting every obſervation, except the meridian altitude for the latitude.   His uncertainty naturally increaſed that of others.   To elucidate this obſervation, I ſhall take the liberty of tranſlating, from the journal of one of our officers, his remark on the land ſeen the 25th.   " Saw land, which ſuppoſe either Copper or Bering's iſland."   I ſhall leave my readers to form their own conjectures.   We were now at very ſhort allowance of water; and the opinion of all hands on board was taken, whether we ſhould ſeek anchorage, and take in a freſh ſupply.   The misfor-tunes of Captain Bering in 1741 were ſo ſtrong in the minds of all the ſailors, that they declared they would rather riſk ſtarving on board than attempt to land on this iſland.

We

We continued a north-weft courfe till the 3d October at noon, when our obferved latitude was 52° 16' 14". By our reckoning, we had paffed Sheeponfkoi Nofs 50 miles, and were only 40 miles from the Bay of Avatfha from our bearings by Captain Cook's chart. The 5th October we got an obfervation of the fun and moon's diftance, at 3° 32' 24" apparent time; which made the longitude 167° 12' 22", latitude 52° 57'; by our reckoning, we were in longitude 157°, nearly acrofs the land of Kamtfhatka.

October the 10th, at noon, the haze clearing a little, difcovered over our ftarboard-quarter Sheponfkoi Nofs, north-eaft 22°, dif-tance 20 miles; latitude obferved, 52° 52' 34"; and fhortly af-ter we faw the mouth of the Bay of Avatfha; but contrary winds and calms prevented our getting into the harbour of St. Peter and St. Paul till the morning of the 14th, at which time numbers of our crew were infected with the fcurvy; but all perfectly re-covered after they had been a few days on fhore. Our Doctor's journal contained the following remarks : " It was only towards " the end of the voyage, when our bread was out, and we were " reduced to a fhort allowance of water, that the fcurvy made its " appearance. At this time peafe and grits boiled to a thick con-" fiftency in a fmall quantity of water, and buttered, were fub-" ftituted for falted provifions. The fymptoms were, coftive-" nefs, a breaking out, with itching, bleeding of the gums " and nofe, pains in the legs, and fome were fwollen. Upon " our arrival, numbers had pains in their joints, with extreme " laffitude, flufhing heat, dry cough, and an oppreffion of the " breaft : bleeding fparingly, thin drink, and frefh fifh, reftored " all hands in a very fhort time.

" The

" The men employed in filling the water cafks at Oonalafhka
" got cramps in their feet and legs, flufhing heat and violent
" head-ache, which was cured by adminiftering fudorifics."

I think this arofe folely from the careleffnefs of the men, in going
with wet feet.   The water was a collection of melted fnow, very
cold.   The weather hazy and damp.   At nights, 2, 3, and 4°
above the freezing point, and at noon only 6, 7, and 8°.

We joined here the reft of our company, who arrived during
the fummer from Ochotfk all in good health.

I fear that my account of our return from the coaft of America
will have feemed tirefome to fome of my readers ; although I have
been as concife as poffible.   One remark, at leaft, I think it neceffary to make; namely, that I am neither failor nor aftronomer;
nor knew aught of either of the fciences until I embarked on the
expedition.

E e   CHAP.

CHAP. XVI.

Mr. Pribuloff appointed to the Sturman's place, vacant by the Death of Mr. Bronnikoff; he goes in search of an Island.—Discovers one which he names St. George's Island, and another, to which he gives the name of St. Paul's.—Arrival and generous Behaviour of an Enemy's Ship, the Mercury, Captain Coxe; and the Astonishment of the Russian Settlers.—The Russian Secretary put in Irons and sent to Irkutsk, on suspicion of improper Correspondence.— Leave St. Peter and St. Paul.—Reach Bering's Island, and narrowly escape a Rocky Point.—Copper Island.—The Islands of Attoo and Agatto.—Semitsh.—Buldyr.—Kyska.—A Cluster of Islands.—Dress and Amusements of the Inhabitants.—Two Natives of these Islands, who had been Attendants on Captain Billings, put on shore.—Leave Tanaga, and after passing several Clusters of Mountainous Islands, arrive at Oonalashka.

OUR first business was, to unload and lay up our ship for the winter; then to dispatch our ship-builder with necessary hands to Neizshni, Kamtshatka, to build a consort for the Slava Russie, to accompany our next year's adventures.

The materials for this purpose arrived with the transport vessel from Ochotsk. The vacant sturman's place, occasioned by the death of Mr. Bronnikoff at Ochotsk, was supplied by Mr. Pribuloff, who accompanied a trader's vessel three years back on the part of Government to collect tribute. At the same time he took

charge

charge of the veſſel as commander, on the part of the trading company; for which he received a ſhare in the profits of the voyage. He made Oonalaſhka, and from his former obſervations that numbers of ſea animals, particularly young kotic, came from the north in the autumn, at the commencement of ſevere weather, he had formed a conjecture, that ſome unknown iſland lay at no great diſtance in that direction; and therefore reſolved, without loſing time, to take on board as many iſlanders as he could obtain, with their ſmall canoes and arms, and be convinced of the certainty or uncertainty of his ſuppoſition.

Twenty-four hours after his departure from the iſland of Oonalaſhka, he diſcovered land. The ſouthern and weſtern parts are ſurrounded by rocks; but the north is eaſy of approach, and affords good anchorage in a commodious bay for ſmall veſſels, not drawing above eight or nine feet water. The whole iſland is volcanic, deſtitute of inhabitants, and only produces the bulbs, plants, and berries, which are to be met with on all the Aleutan iſlands. They found the low lands and the ſurrounding rocks covered with ſea animals, particularly the urſine ſeal (kotic), and ſea-lion (ſivutſha); and with the ſkins of theſe animals they nearly loaded their veſſel. Pribuloff called this St. George's Iſland; and obſerving another iſland to the north, at the diſtance of 44 miles, he went thither in a large baidar, accompanied by a number of Aleutes. This iſland is much ſmaller than that of St. George, and he named it St. Paul's: this, as well as the former, was the retreat of immenſe herds of ſeals. On the iſland of St. George they paſſed the winter, and found the inland parts overrun with foxes, which afforded them a profitable chaſe. It alſo abounded with the tuſks of the walroſs, which they picked up on the ſhores.

E e 2

Laſt

Laſt autumn he returned to Oonalaſhka, where he paſſed the winter. A European veſſel put into the bay of Udagha, which Mr. Pribuloff viſited: it was the Mercury, Captain Coxe, copper-bottomed, and mounting ſixteen guns. From this veſſel, which he ſaid had only two maſts, he received intelligence of the war between Ruſſia and Sweden. The Captain was inquiſitive about the Ruſſian eſtabliſhments, their force, and ſhipping: to explain which, Pribuloff took the Captain and his officers to their habitations, but could not treat them with any thing except ſaranna, berries, the dried meat of the ſea-lion, and fiſh, without bread. They expreſſed aſtoniſhment at every thing they ſaw, but moſt at their manner of living, &c. On their return to their own ſhip, they ſent Pribuloff a ſupply of bread, brandy, and other neceſſaries, ſome articles of dreſs, and a quadrant, as preſents; and a few days after left the iſland.

Nothing in the world can aſtoniſh a Ruſſian more than diſintereſted liberality, or any kindneſs without ſome proſpect of future benefit. Greatneſs of ſoul is applied to every man who is juſt, and grants his ſervants ſome few indulgences; every thing beyond this is called folly, and is ſure to be impoſed upon: nor have they any ſentiment of feeling, except it be excited by blows. Taking this for the ruling character of the Ruſſian hunters, it will be eaſy to conceive the aſtoniſhment of Pribuloff and his companions at the liberality of Mr. Coxe: but how much was their amazement increaſed, when, on their returning to Ochotſk, they were informed, that this very Captain commanded an enemy's ſhip, and actually had a Swediſh commiſſion to deſtroy the Ruſſian eſtabliſhments! They could not imagine what inducement he could have to ſhew them any mercy, much leſs to heap kind-

neſſes

neffes upon them. Pribuloff himfelf faid, " They had every thing, " and faw that we had nothing worth their taking; therefore " they made us prefents; for they were afhamed to be enemies " to fuch poor wretches *."

I am inclined to think that his conjectures were juft; and I feel myfelf interefted in relating this anecdote, which, in my opinion, does fo much credit to an European failor, of whatever nation he may be.

Nothing material happened this winter, the greateft part of which we paffed at Bolfhoiretfk, receiving frequent intelligence of the progrefs at Neizfhni under the direction of Captain Hall, who acquainted us that his veffel would be ready to put to fea as foon as the river Kamtfhatka fhould be free of ice. One circum-ftance, however, I think it neceffary to mention. The Ruffian fecretary, Vaffiley Diakonoff, having given diffatisfaction to Cap-tain Billings, and being thought to have entered into a private cor-refpondence with Mr. Shelikoff, and difclofed fome fecrets of the expedition, was put in irons and fent to Irkutfk, to anfwer for his conduct on the return of the expedition.

I forbear making any comment upon this bufinefs, or giving any particular account of Kamtfhatka, until I take my final de-parture from it.

* The liberality with which the Expedition under Captain Cook treated the natives of every place they touched at, infufed into their minds an aftonifhing idea of the wealth and profufion of the nation from whence they came: this, therefore, was a very natural conclu-fion of Pribuloff.

Coreilin was the commander of the hunting parties on fhore; and he alfo received feveral prefents from the Mercury.

We

We paffed the winter in excurfions of pleafure, and in dancing and card parties, chiefly at Bolfhoiretfk, where the luxuries of life are more plentiful than in the harbour of St. Peter and St. Paul. The froft was fometimes very fevere, and we had for a few hours 21° below the freezing point of Reaumur. Two or three earthquakes happened about the neighbourhood of Neizfhni; but, except a flight fhock on the 21ft November at noon, none of them reached the harbour of St. Peter and St. Paul.

In the beginning of the month of April 1791, all hands repaired to the harbour. Inftructions were fent to Captain Hall (to whom the command of the fecond veffel was allotted as fenior officer), to be at Bering's ifland by the 25th May; and if he did not find us, to wait till the 30th: we alfo were to wait till the 30th for his veffel, if we arrived earlier. In cafe we fhould not meet there at all, the fecond place of rendezvous was appointed at Oonalafhka.

We took in a good fupply of water, rolling the cafks over the ice of the harbour to the fhip. We alfo took a greater quantity of provifions than in the preceding voyage; though the falted meat was lefs nourifhing, having lain fo much longer in the cafks. The different meffes took a good ftock of dried and pickled falmon, berries, wild onions, &c.

By the 1ft of May the bay of Avatfka was clear of ice, and not before, owing to the feverity of the winter: but the inner har-bour of St. Peter and St. Paul remained frozen up.

2

On

On the 8th, we broke the ice of the harbour to make a paſſage for the ſhip, and hauled her into the bay. The thermometer ſtood at 2° to 4° of heat, the wind blowing right againſt us till the 13th, when it fell calm. We now took the ſmithy and all hands on board, and hauled off the battery point. Baffling light airs detained us till Friday the 16th, at four o'clock in the morning, when we weighed anchor with a gentle breeze from the north, and ſaluted the battery with ſeven guns, which was returned with an equal number. The wind falling ſcant, and ſhifting to ſouth-weſt, with a contrary current, we made but little headway, and caſt anchor at four P. M. in the mouth of the bay, the lighthouſe bearing north-eaſt 86°, diſtant about one mile. The next morning a moderate breeze ſprung up from the eaſt, which brought with it a very thick fog. The tide ſetting againſt us to the weſt, at three knots and a half, our ſhip drove unobſerved, and we diſcovered that we had a flat ſtone bottom at twelve fathom. We drove very near the ſouth-weſt rocky ſhore, ſent a kedge to the north-eaſt into good anchorage, weighed anchor, and hauled a-head. At ten A. M. a gentle breeze ſpringing up from ſouth, we took in our kedge, and ſtood about two knots north north-weſt, when, coming to a good bottom, we brought up with our beſt bower. At noon we obſerved the latitude 52° 55′ 32″.

All the 18th we had variable light airs, with cloudy and hazy weather. The Kamtſhadals that were out among the rocky iſlands, ſeeking eggs, brought us a very conſiderable ſupply, as alſo of ſea-fowl.

Monday, the 19th May, we weighed anchor at four A. M. with a gentle weſt ſouth-weſt breeze, ſtood out of the bay of
Avatſha,

Avatfha, and hoifted our boats on board, going eaft by north at three knots. At noon, our bearings were Povorotnoi Muis (Cook's Cape Gaveria), fouth-weft 23° 30'; Villuitfhefkoi Peak (Paratounka Sopka) fouth-weft 72°; lighthoufe, by true compafs, north-weft 58° 23', diftant feven miles and a half; latitude of fhip's place obferved 52° 49', longitude 158° 56', variation one half point eaft, from whence we took our departure, and continued our courfe all day.

On the 20th we faw immenfe numbers of grampuffes, porpoifes, and many whales. We had a frefh gale from the north, hauled the wind, and kept a courfe eaft north-eaft, under clofe-reefed top-fails. In the afternoon, moderate wind. The 21ft, at noon, latitude 53° 9'; longitude 161° 39'. Sheponfkoi Cape bore due weft, diftant about fifteen leagues. I make this cape in latitude 53° 9', longitude 160° 3', variation three-fourths of a point eaft. Variable light winds, and calms, prevented our making much head-way till Saturday the 24th; when, early in the morning, a breeze fprung up from the fouth-weft, and we failed north at the rate of fix knots. At noon our obferved latitude was 54° 14', corrected longitude 162° 30'. The eaftern extremity of Kronotfkoi Cape bore north 2° 30' weft. In the evening the breeze died away, and calms and light airs followed till the morning of the 27th, when we had a gentle breeze from the weft fouth-weft. At noon our latitude obferved was 54° 45' 22", longitude 165° 36'. At three P. M. faw Bering's ifland, the fouth-weft point of which bore north-eaft, the fouth-eaft extremity north-eaft 73°. At eight P. M. the wind frefhened, with hazy weather, and fqually; the land was about four miles to leeward, and a detached rock off the north-weft extremity a-head of us. Mr. Bakoff, who had the watch, firft difcovered the dangerous

6

fituation;

fituation ; and it was owing to his prefence of mind, in immediately crouding all the fail the fhip could carry, that we weathered
this rock, at not the fhip's length from it, carrying her gunwale
nearly under water. Having cleared this point at eleven P. M.
we ftood away more large, with very hazy weather.

This ifland's fouth extremity bears by true compafs from the
harbour of St. Peter and St. Paul north-eaft 67°, its diftance 192
miles, trending north-weft 35°, forty miles. The weft fide of the
ifland is mountainous, and covered with fnow; the fummits were
hid in the haze and fog. The north point is low land, free of
fnow. Here are two bays where merchants' galliots winter; but
they are fhoal, dangerous of approach, and expofed to the north
winds. A fmall rivulet runs into each of them, in which tranfparent white pebbles are found; and fometimes, after a hard
gale of wind from the north, fmall pieces of native copper are
caft on the fhores. The north point is in latitude 55° 25', longitude 166° 15'.

The 28th, at noon, our latitude was 55° 14' 23", corrected
longitude 166° 50'. At two P. M. the fun and moon's diftance
gave the longitude 166° 52' 45", which perfectly agreed with our
fhip's reckoning: variation one point eaft. The wind blowing
frefh from the fouth-weft, we could not attempt to enter the bay
to feek for the fecond veffel; and it was refolved to profecute our
voyage to Oonalafhka.

The 29th, at three o'clock in the morning, being very foggy,
we faw Copper ifland aftern of us; fo that we muft have paffed
it very clofe indeed. Our courfe was eaft fouth-eaft, the wind
F f   blowing

blowing fresh from the south-west. Owing to thick weather, we could only observe that Copper island is mountainous, bearing from Bering's island's south point north-east 65°, distant 27 miles, trending south-east 61°, twenty-five miles: rocks between the islands, and off their northern extremities. At five P. M. the sun and moon's distance proved our longitude 169° 0′ 15″, latitude 54° 14′. In the evening the wind died away.

The 30th, our latitude at noon was 53° 43′, longitude 170° 12′. At seven P. M. we saw land, a lofty mountain covered with snow, south-east 30°, which was soon hid by the haze and darkness of the night. Light airs all night and the next day, with a considerable swell from the south. Hazy weather prevented our seeing the land again till Wednesday the 4th June at three A. M. when the west extremity of Attoo bore north, distant 13 miles, and the eastern point north-east 72°. Going east north-east at two knots, with little wind from the west-north-west, at noon latitude by account 52° 32′, longitude 172° 15′, variation one and one-fourth point east. At four P. M. the west of Attoo bore north-west 31° 30′, Agattoo's north-west point, north-east 60°. We threw the ship in the wind, and got soundings with 75 fathoms, stony bottom.

The island of Attoo is mountainous and covered with snow. Its western end bears by true compass from the south of Bering's island south 61° east, 215 miles distant: its direction east and west about 60 miles. Detached rocks are off the west point; and its south side has several openings appearing like coves, but exposed to the south. From the east of Attoo to the west of Agattoo the distance is 20 miles south-east one-fourth east, trend-

ing

ing eaft about 16 miles. Here alfo appear fome openings, but the entrances are barred by a reef of rocks. The weft extremity is low land gradually afcending. Eight miles from the point is a very lofty mountain, the top of which was hid in the clouds, as was alfo the higher land towards the eaft. Ten miles north is the little ifland of Semitfh, and off the eaftern point is a ftill fmaller ifland.

We kept an eafterly courfe; and at noon of the 5th our obferved latitude was 52° 10' 25", correcled longitude 174° 17', with a very hazy horizon. At two P. M. we faw Buldyr, northeaft by north one-half eaft: fhortly after we had rainy and thick weather. Buldyr bears by true compafs from the ifland of Agattoo north-eaft 88°, diftant 70 miles. This is an oval rock, very lofty, fix miles from north to fouth, and four miles acrofs. Off the eaft and weft points are detached rocks, to the weft they extend to a confiderable diftance.

The 6th, at three o'clock in the morning, we faw the ifland of Kyfka to the fouth-eaft, a detached rock fouth-eaft 64°, and fhaped our courfe through the ftraits, to get to the north of the iflands.

Thefe ftraits are 64 miles wide. The north point of Kyfka bears eaft from the fouth point of Buldyr; its direction fouth by eaft, and extent 26 miles, terminating in a point of moderate height, and 20 miles acrofs in the wideft part: there is fome low land about the eaft extremity, and it contains many rocks. At noon our latitude by account was 52° 23' 20", longitude 177°, when we rofe a clufter of iflands; the moft weftern of which

The Island Attoo bearing North distance 13 Miles.

The Island Agattoo bearing N.E.

The Island Buldyr bearing North distant 8 Miles.

The South-east side of the Island Kyska.

Published Feb. 23.ᵈ 1802, by Cadell & Davies Strand.

J. J. Wᵃⁱᵏᵉʳ Sculp. 102, Strand.

is called Sigoola, 14 miles eaft of Kyfka. This is nearly round, and nine miles in circumference. Kriffey ifland is about an equal diftance from Kyfka to the fouth-eaft, fmall and rocky. The ifland of Amtfhitka lies about eight miles eaft north-eaft of Kriffey, and trends eaft nearly 60 miles in extent. It has an expofed bay to the fouth ; the north fide is acceffible for boats ; but the clufter of iflands on this fide render its approach by veffels impoffible. Off its eaftern extremity, due north, at the diftance of 28 miles, is the ifland Semi Sopefhnoi, or Seven Peaks, trend-ing eaft and weft 22 miles. The fog, however, foon hid thefe iflands from our fight.

Thick weather prevented our feeing land till the 9th, when, at noon, the fun broke through the haze, and we fuddenly faw the land over the ftarboard fore-yard arm, appearing clofe to the fhip : a tremendous barren mountain ftreaked with fnow imme-diately difcovered its bafe, bearing from north-eaft 62° to fouth-eaft 46°, diftant in the neareft place about half a mile (but no foundings with 100 fathom line) ; a perpendicular rock. The fun's altitude in the haze was 61° 10', which gave the latitude 52° 5' 21" ; by fhip's reckoning 52° 6' ; longitude 180° 22' ; varia-tion one point and a half eaft. The wind blew pretty frefh from the north ; and, as it was impoffible to weather its point, we fhaped our courfe fouth-eaft by fouth. At firft we took the rock for the Volcano Gorelloi ; but foon difcovered it to be the north-weft extremity of Tanaga, which is formed by an uneven-topped volcano, appearing like a clufter of mountains. One terminates in a conical point, of extreme height, emitting fometimes a co-lumn of fmoke. They are all covered with fnow, which def-cends in ridges to below the middle of the mountain, but much

darkened

darkened by the quantity of afhes upon it. This mountain occupies a fpace of eight miles fouth, and fix miles eaft by north.
South fouth-weft eight miles from the north weft extremity of the
ifland, the high land terminates by a projecting rocky cape, fhar-
pened by feveral detached needle rocks, behind which we thought
there might be good anchorage. Captain Saretfheff volunteered
to explore, and went in the evening in the long-boat with this
intent. We ftood off and on to wait his return ; a thick fog,
however, fell upon us, which continued till the next day at noon.
During this time we very frequently experienced ftrong rippling
tides in various directions, but chiefly fouth and fouth-eaft. Our
diftance was about one mile from fhore, and we faw the long-
boat pulling on board, which foon arrived. Mr. Saretfheff found
pretty good anchorage fix miles fouth fouth-weft behind the
needle rocks ; our obferved latitude at noon was 51° 56′ 3″, one
mile eaft of the neareft land. It falling calm, we towed into the
bay, and came to anchor at fix P. M. ; Gorelloi Volcano bore
weft north-weft, diftant 22 miles. In the fog yefterday we muft
have paffed this mountain very clofe indeed. We did not fee the
top of this volcano on account of the clouds refting upon it ; but
the fhores are very fteep, and there is no accefs, except in very
calm weather, on the fouth-weft part for boats ; its bearing from
Kriffey ifland fouth 81° eaft, diftant 107 miles, fix miles from
north to fouth, and three miles from eaft to weft. I have defcribed
the north-weft part of Tanaga to the fpiral rocks trending fouth
fouth-weft eight miles ; thefe rocks form the north bounds of a
fmall bafin, in which we came to anchor in latitude 51° 52′, and
longitude 180° 25′. It is about two miles and a half in circumference, with a fandy bottom feven fathom ; and, at the head of the
inlet, is a very convenient watering-place ; it is, however, expofed
to the north-weft winds.

From

From this place the land trends weft by fouth eight miles, low and very verdant; terminating in a fandy cape, from whence the ifland ftretches away fouth by weft 15 miles; all, except the north, is low land, with frefh water lakes; but interfperfed with fome rifing grounds, near which are the defolated dwellings of the former inhabitants. We found the earthen habitations in one place contain about 20 women, and only a few men, either old or very infirm, which conftitutes the prefent population of the ifland, exclufive of a few children. The male inhabitants had been taken by Luchanin's company of hunters to affift them in the chafe; and what induced the remainder to fuppofe that it was not the intention of Luchanin that they fhould ever return, but form an eftablifhment perhaps on the coaft of America (which they call Kanaifki Land), was, that he had alfo taken as many women with him as he could poffibly ftow away in his galliot. We learnt, however, from thefe remaining inhabitants, that their companions did not go voluntarily. This ifland was formerly very well inhabited; but the Ruffians have almoft depopulated it, which is completely the cafe with thofe to the weft.

The inhabitants drefs exactly like thofe at Oonalafhka; but the women have not fo many ornaments. They fpeak different dialects of the fame language as at the above-mentioned ifland. Their dances and diverfions, however, feem different. They are more graceful in their motion, extremely modeft in all their actions; and quite unlike all other favages that I have feen, by being free from lafcivioufnefs. Young men amufe themfelves with jumping on the fkin of a large fea-lion, held in the air by four or fix men. They leap and lighten upon their feet, and by degrees are thrown up to an immenfe height: when they are tired they leap off upon the ground. I attempted to leap in this manner,

I

but

but could not fucceed; for the fudden jerk either caufed my knees to bend, or elfe threw me out of the centre; and they explained the caufe by telling me, that I looked upon the fkin, whereas I ought to keep my body erect, and look upwards; at the fame time I fhould not leap, but let the men throw me up. Their boats are larger and more heavy than thofe of Oonalafhka, though made upon the fame principle.

We had on board three natives of this and the neighbouring ifland of Kanaga, taken from hence in the year 1785 by Gregory Shelikoff, of whofe behaviour upon thefe iflands we received very unfavourable accounts. Two of them had been the attendants of Captain Billings from the time of his firft arrival at Ochotfk in 1786; and now embarked with a view of being left at their native habitations; to which, notwithftanding the defolation that they beheld, they flew with fatisfaction; (a ftrong proof of the attachment of mankind to the country where they have paffed the years of innocence and happinefs!) content in the poffeffion of a piece of paper which exempted them from the flavifh demands of the Ruffians * (in cafe they choofe to pay any attention to it). Thus rewarded, with the addition of a few articles of drefs, the free gift of different officers on board, and with a very fmall quantity of tobacco, they were put on fhore.

This was not quite the ftile in which Omai was returned to his family and friends by Captain Cook; for all the wealth that thefe poffeffed between them could not create envy among their brethren; nor could all the accomplifhments which they had ac-

* I hope that my readers will not confound the character of thefe defperate exiles with the general character of the Ruffians, who are kind and hofpitable to an excefs.

quired,

quired, during a fix year's fervice in conftant employment with the utmoft diligence and fidelity *, prove any recommendation to their relatives, or qualify them to obtain their living by the productions of the chafe; for only one of them was capable of rowing in their fmall baidars, and I believe that only in fmooth water. I cannot fee any other means of their fupporting themfelves than by digging the edible roots, and obtaining the fhell-fifh with which the fandy fhores abound, particularly cockles of an extraordinary fize.

The rocks have alfo a variety of mufcles, and feveral fpecies of limpets that adhere to them juft at low water mark; a particular fpecies in great abundance called by the Ruffian hunters baidars, from the great refemblance which they bear to their open boats, with a row of jointed fhells along the centre of the back; thefe are devoured by the natives both raw and dreffed, and I thought them very good eating; the largeft were about three inches long, and one inch broad, very flefhy and firm. Whales are frequently caft afhore upon the fandy point of this ifland, and afford food and light for a confiderable time.

One fpecies of whale is frequently caft on fhore both on thefe iflands and on the coaft of Kamtfhatka, which the natives never eat, but only ufe the fat to burn. They know no difference in its appearance; but obferve that neither gulls, nor any bird of prey, or fox, will eat of it. They fay, that the Ruffian hunters

* One of thefe lads attended Mr. Main and me at the time we were at the Kovima. Main afked him, what the favages would do fhould he (Main) fall into their hands? The boy replied: " Sir, you fhall never fall into their hands if I am with you; for I do always carry a fharp knife about me; and when I fee that there is no poffibility of your efcaping, I will ftab you to the heart, and then they will not meddle with you."

have

have ufed it for food; that its fat turns in the ftomach to an oil of fo fubtile a nature, as to pafs through all the pores of the body, while the flefhy parts are emitted in an undigefted ftate; and that if thofe who have eaten it have formerly had wounds or ulcers, although thefe have been cured for years, they break out afrefh. Several of the hunters told me, that they had eaten of this whale, and that the account which the natives gave of the fubtilenefs of the fat, and the undigefted ftate in which the more fubftantial parts paffed through them, was true; and that fome of their companions, who had been cured of the venereal diftemper, became again violently affected with that difmal difeafe, merely from this food. The fame property, however, is attributed to the flefh of whales in general.

This was the only ifland on which we obferved the eider-duck; and it was about the lakes here in great plenty. The dreffes made of their fkins are efteemed the beft of all the feathered tribe, being more foft, warm, and ftrong, than any others.

We remained at anchor till Sunday the 15th, and the whole time experienced hazy and mifty weather, which prevented our feeing the fmall rocky iflands to the weft, which were formerly the places of refort of the fea-otter and other marine animals, now nearly extirpated, or entirely driven from thefe parts. At three P.M. we got under weigh, with a gentle breeze from the fouth-eaft, and ftood away weft north-weft to double the north cape of the ifland; but, the wind being fcant and fhifting to the north-eaft by eaft at fix P.M., we kept working to windward all the 16th.

G g   The

The 17th, ftill plying, at noon our obferved latitude was 52° 7' 55". We made the ifland of Kanaga feven miles diftant from that of Tanaga, and faw the fmoke afcending from the hot fpring at the foot of an extinguifhed volcano on the ifland of Kanaga, off which at 12 miles we alfo obferved the fmall ifland called Bobrovoi, from the number of fea-otters that formerly held their refting-place upon it.

The wind ftill continuing from the north-eaft quarter, with very thick weather, which prevented our difcovering the leeward iflands, we refolved upon paffing the ftraits weft of Tanaga, where we ftood at anchor to get to the fouth of the iflands. At three A. M. of the 18th, we ftood fouth-weft by fouth, with a gentle breeze from north-eaft by north, which frefhened by noon.to a brifk gale. The weather was hazy; our latitude by account 51° 48' 5", the body of Gorelloi Peak bearing fouth-weft 72°, when we rofe the low ifland of Hluk fouth-weft 6°, diftant from the low fouth-weft extremity of Tanaga 12 miles. We fhortly after faw the rocky iflands between Illuk and Gorelloi. The 19th, eafterly fqualls and thick weather, with flying clouds at noon, when we got a fight of the fun; and the latitude obferved was 51° 27' 20", correfted longitude 181° 29', variation of the com-pafs one point and a half eaft. Afternoon rainy with a hard gale. At five P. M. it brought us under our courfes clofe hauled on the ftarboard tack.

The 20th, at noon, having moderate wind from the fouth, fet clofe-reefed topfails, and ftood away eaft by north, latitude by ac-count 51° 2', hazy weather, and no land feen. The 21ft, at noon, obferved latitude 51° 4' 57", correfted longitude 182° 22'. At

five

five P. M. faw the ifland Adach; its weftern low extremity north-weft 15°. At feven, having made the laft two hours feven knots and a half north north-eaft, the low extremity of Adach bore north-weft 39°, diftant 17 miles, and we rofe another ifland north north-weft, when the haze hid all land until Sunday the 22d, at feven A. M. when we faw mountains, and at eight found ourfelves near a clufter of fifteen fmall iflands, mountainous and of various forms. At noon our latitude obferved was 51° 58' 38", correted longitude 184° 48'. We were now failing with a moderate breeze from the fouth to the north-eaft, at the rate of fix and feven knots through the ftraits formed by the clufter of moun-tainous iflands before mentioned; fome of them not half a mile off (foundings from 30 to 50 fathoms, fhells and coral); the largeft, Gorelloi ifland (not Peak), north-weft 57°. My me-moranda, taken from the original log, not being here very dif-tinct, I am unable to give the exact bearings and diftances of thefe iflands. We paffed the north point of Alcha, however, about four miles from the land, when we were furrounded by a thick fog.

The 23d, at noon, our latitude by account 53° 4', longitude 187° 48', we allowed variation one point and three quarters eaft; very foggy all day. Continued our courfe north-eaft half eaft 24 hours, until Tuefday the 24th at noon, when our diftance run proved 128 miles north eaft 80° 24', making our latitude 53° 27', longitude 191° 28', rainy, mifty, and foggy weather, which prevented our feeing land till half paft one, when we fuddenly faw land fouth-eaft, about two miles and a half diftant, which was known to be a promontory on Oomnak; upon which we fhaped our courfe north-eaft. At five P. M. we rofe the north-weft ex-tremity of Oonalafhka, fouth-eaft 85°. At nine P. M. paffed the

remark-

remarkable rock, refembling a fhip under fail, in the middle of the ftraits, between Oomnak and Oonalafhka, and opened Tfherneffki bay fouth-eaft 16°. Upon feeing land ahead, hauled the wind and ftood away north. At ten P. M. being about three miles from the fhore of Oonalafhka, we fent the baidar to examine the coaft, which was hid from our fight in the haze, and ftood off- and-on all night. The next morning, Wednefday, the 25th June, at four A. M. with a gentle breeze from the weft, being well in with the land, feveral natives came on board, under whofe pilot- age we ftood into the bay of Amoknak; and at three P. M. came to anchor in the bafin of Illuluk, about 20 fathom from the dwell- ings that bear this name, and fhortly after fent the obfervatory on fhore.

Captain Hall was not yet arrived: a circumftance rather unac- countable to us, except on the fuppofition that he could not get fo foon ready to leave Kamtfhatka as he expected.

CHAP. XVII.

Captain Billings abandons all thought of re-visiting the American Coast to the south of Cook's River, and prepares to sail for the Bay of St. Laurence.—Reflections of the Author.—Sail for the Bay of St. Laurence.—The Islands of St. George and St. Paul.— Gore's Island, &c.—Captain Billings, &c. land on the Continent of America; of which visit some Particulars are given from the Memoranda of a Gentleman in the Party.—Come to anchor in the Bay of St. Laurence.

CAPTAIN BILLINGS now declared, that he was resolved to abandon every idea of revisiting the American coast to the south of Cook's River; but determined to proceed (so soon as he had taken in a fresh supply of water, landed provision for the vessel under the command of Captain Hall, and taken on board ballast in its stead) direct to the Bay of St. Laurence, in the land of the Tshutski, where two petty officers, Dauerkin and Kobeleff, sent from Ochotsk in 1789, had orders to wait our arrival; and, in case Captain Hall should not arrive in the mean time, orders were to be left with Mr. Allegretti (his surgeon), Ivan Alexeeff (an ensign), and one sailor, who were to remain on shore to guard the provision, that he should immediately follow us to the above-mentioned Bay of St. Laurence, where Captain Billings meant to land, without even attempting to see how far he might be able to pass through Bering's Straits; asserting, that the season was too far advanced, and that he should have an opportunity of ascertaining every thing necessary by land.

Nothing

Nothing in the world could have afforded me lefs fatisfaction than this refolution, which I regarded as the conclufion of an expedition that was fet on foot with unbounded liberality by the moft magnanimous fovereign in the world; which had raifed the expectation of all nations to the higheft pitch, and induced mankind to anticipate the fatisfaction of obtaining the moft complete knowledge of the geography of this unknown part of the globe, together with a conviction of the exiftence or non-exiftence of a north-weft paffage. But, alas! after fo many years of danger and fatigue; after putting the government to fuch an extraordinary expence; after having advanced fo far in the attempt, even at the very time when we were in hourly expectation of our confort, and, as appeared to me, being juft entering upon the grand part of the undertaking, thus to abandon it, was the moft unaccountable and unjuftifiable of actions.

I defpaired of feeing Captain Hall again, at leaft until our return to Kamtfhatka, or perhaps St. Peterfburg, unlefs we fhould be fo fortunate as to join company before leaving this ifland, which might, perhaps, alter the prefent plan, and lead us to purfue the real object of the expedition.

The remonftrances of Captain Saretfheff at the Kovima, on the Icy Sea, &c. &c. and in fact the reprefentations of every officer who had hitherto prefumed to have an opinion, were always treated by the Commander with petulant and illiberal retorts. I have, indeed, had too frequent opportunities of obferving, that rank and power intoxicate the poffeffor, unlefs they have been the reward of real merit, or the confequences of feniority in actual fervice; in which cafes, the value of authority is known, as wealth gained by labour,

10

and

and not ufed as the accidental and unexpected inheritance of a prodigal.

Excepting Captain Billings, Mr. Saretfheff was the only naval officer on board; and I can affirm, that the latter was the only fcientific navigator in our Expedition: a gentleman, who poffeffed that particular modefty which is always the companion of merit, with feelings the moft acute, refined by true fentiments of honour; to which (at one time, at leaft) he had hopes of adding fome luftre in the prefent undertaking. His duty at length got the better of his feelings fo far, as to lead him to afk, whether no other perfon could be fent by land, while Captain Billings himfelf made a fecond attempt by fea? And, whether it was abfolutely neceffary for him (Billings) to go? Receiving only evafive anfwers, however, he entertained hopes of better fuccefs if Captain Hall's arrival fhould ftrengthen his efforts *.

* I would moft willingly have drawn a veil over this part of my narrative; but that my fo doing, I thought, would have been more unjuft than the caufe that gave rife to it, and at the fame time would have eclipfed the merit of other officers on board. The officers of the three watches were, Captain Saretfheff; Mr. Bakoff, a gentleman whofe bufinefs it was to take care of the fhip's materials, boats, &c. but who knew nothing of navigation or numbers; and Mr. Batakoff, a fturman, or mafter, whofe duty was to keep the log reckoning, con to the helmfman, &c. but who had nothing to do with the working of the fhip; his learning extended to a common day's work; taking the fun's altitude at noon, and its azimuth for the variation; making furveys, &c. Thefe gentlemen, from their experience, were well qualified to keep watch; and, although not the moft learned of men, poffeffed

" Good fenfe, which only is the gift of Heav'n,
" And, though no fcience, fairly worth the fev'n.

They were active, zealous, and enterprifing; particularly Mr. Bakoff, who was alfo bleffed with aftonifhing prefence of mind in all cafes of neceffity or danger. The failors and petty officers were divided into two watches: and it may be proper here to remark, that not one of the common failors had ever feen a fhip before; which, indeed, was the cafe with all the petty officers, except three.

Having

Having landed the provisions and stores for the second vessel, taken in a sufficient quantity of ballast, and a supply of fresh water, we were completely ready for sea on Monday the 7th July. In the evening Mr. Allegretti (Captain Hall's surgeon), Ivan Alexceff (an ensign of jagers), and a sailor, were put on shore, to guard the stores, and with instructions for Captain Hall to follow us to the Bay of St. Laurence. At nine P. M. weighed anchor; but, falling calm, we hauled about two knots out of the basin, and again brought up.

Thursday July 8th, weighed at four A. M. with a gentle south-west breeze; but shortly after a calm compelled us a second time to come to anchor in 18 fathoms in the Bay of Amoknak, over a fine sandy bottom. At two P. M. a gentle breeze sprang up from the south south-east; which freshening, at three we got under weigh, and stood out of the Bay of Amoknak, at the mouth of which the ship's latitude, by bearings from the observations on shore, was 54° 8', longitude 193° 17' east from Greenwich; variation of the compass two points east; the volcano on Acutan bearing north-east 62°, Oonalgi south-east 73°. We soon after saw Akoona to the north-east. One of the natives here overtook us, and, wishing to accompany us, was taken on board with his small baidar. At eight P. M. stood away north-west, with a fresh south-east breeze and foggy weather. At midnight the wind veered to north-west, and soon after increased to a gale, with flying clouds. At noon, our observed latitude was 54° 59' 38", corrected longitude 193° 1'. In the afternoon the wind became more moderate.

By noon of the 10th we had run 29 miles only, north-west 36° 57'. In the fore-part of the day, I begged Mr. Saretsheff to

heave

heave the lead, upon a suppofition that he would get foundings, which he did at 80 fathoms, mud and fand, and in the evening at 75 fathoms. It had been very hazy all day, and continued fo all the next day, with little wind at north north-weft, and weft by north. At noon of the 11th we faw feveral herds of fea-lions fporting, fea-birds, and weeds floating. Our 24 hours run was north-weft 49° 20, 75 miles. Foggy and mifty weather continued all the afternoon. At night, having a frefh breeze fouth by weft, we clofe-reefed our top-fails, and, in hopes of feeing in the morning the iflands difcovered by Prebuiloff, laid-to with the main-top-fail to the maft. Saturday the 12th, early in the morning, we righted fails, ftood to weft north-weft, and faw land in the fog, bearing north north-weft. At noon our latitude, by account, was 56° 59', longitude 189° 45', when the fouth extremity of St. George's Ifland bore fouth-eaft 57°, diftant 16 miles; and foon after we faw the ifland of St. Paul. Thefe iflands appear hilly, though not mountainous; many vallies are difcernible, covered with green plants, fuch as are to be met with on all the Aleutan iflands; but there is not a tree or a fhrub upon the ifland, except fome low berry-bearing bufhes from 12 to 16 inches high. A reef of rocks off the fouth and fouth-weft fides of the iflands ex-tends about three miles; fome of them are difcovered by the breakers, while others are confiderably above water. They are not furnifhed with any harbour; but to the north-eaft are bays fhallow and expofed, which, however, bad as they are, afford a landing place to the hunters, whofe firft bufinefs is, to fecure their veffels by hauling them on fhore. The weft extremity of the ifland of St. George bears, by true compafs, from the north point of Oonalafhka north 39° weft, diftant 190 miles, trending eaft by north, one-half eaft 19 miles, and is about eight miles wide. Luchanin's company are now here; but by the accounts of the

H h   native

native on board, and of others with whom we spoke at Oonalashka, they get but few animals. Drift wood is also scarce, which was plenty at the time when Pribuiloff first discovered these islands. This company of hunters have also a few hands with them from Oonalashka.

At eight P. M. the fog hid the land: we had a moderate breeze from the east south-east, and our course was north-west one-half west, allowing two points; variation east.

Sunday the 13th July, in the fore-part of the day, we had fresh wind east south-east, and were going under an easy sail north-west at six knots. At noon very hazy; latitude by account 58° 38', longitude 188° 28'. The afternoon was hazy and misty; and at four P. M. we had a moderate south wind, which veered at seven P.M. to south-west. We were steering north-west and by north, but, on account of night approaching, slackened sail.

On the 14th we had a moderate south south-west breeze, with hazy and foggy weather, and kept our north-west course. At seven A. M. we saw in the haze land to the north-east, which we soon after discovered to be Cook's Pinnacle Island, and stood in for it north by east. At eight, the wind veering to south south-east, we saw a rock in the fog right a-head, so wore ship, and steered two knots south-west one-half west, when we again resumed our former course. At ten A. M. saw Gore's island a-head, and soon after observed that it extended considerably west of the north. At noon, the meridian altitude in the haze gave the latitude 60° 30' 50", longitude corrected 187° 15'. Our distance was now about two miles from the south-east extremity of the land, trending north-west 61°. We ranged along the whole of

I the

a

S.E. 48 30

Gores Island —

— drawn while at Anchor in the Straits.

The Detached Island 4 Miles W by N from Gores Island

The Island Semisophosmoi as it appeared the 7 June 1792 bearing West dist 5 Miles.

Published Feb. 23 1801, by Cadell & Davies Strand.

the fouth-weft fide of the ifland at the diftance above mentioned; and obferved, that the land was moderately high, and that fnow lay upon many of the higher mountains, the fummits of which were hid in the fog. There were many bays, backed by low land, and fome of them may, probably, afford good anchorage. The capes confifted of projecting promontories, with detached rocks extending out from 50 to 100 fathom. The vallies appeared very verdant; but the high land was barren and rocky. The extent of the ifland is 26 miles. Pinnacle Ifland lies due fouth at the diftance of eight miles, and is a remarkably barren rock, replete with lofty pinnacles, like ftacks of chimneys, with detached rocks off it in every direction.

At the diftance of four miles from the fouth-weft extremity of Gore's ifland, in a direction weft by north, is another rocky ifland trending fix miles north north-weft. At four P. M. we entered thefe ftraits, with foundings at 12 fathom, over a fine fandy bottom. With a view of feeing whether the iflands were inhabited or not, we came to anchor in the mid-channel, lowered our jolly-boat and baidar, and went afhore on the weftern ifland. We found a good landing-place in a fmall bight behind a detached rock, which bore due weft from the veffel; the beach extending about 10 fathom from the perpendicular rocky fhore, covered with drift-wood, the bones and tufks of the walrofs or morzfh, the bones of whales, the back-bone, with ribs adhering to it, of fome large animal (I fuppofe the white bear), and fragments of rocks; agates, and other pebbles, &c. The compofition of the ifland feemed to be mountains of jafper, fome green and red, but in general yellow, veined with tranfparent ftone like calcedoni. I afcended one of the narrow chafms in the rock to the top, which I found level, covered with mofs, and fome fuch low plants as I

H h 2   had

had feen on the borders of the Icy Sea; foxes were numerous, of the black, red, and blue (or arctic) fpecies. There appeared to me to be no earth upon the ifland, except the dung of animals, and of myriads of fea-birds, whofe fhrill notes almoft prevented our hearing each other fpeak: thefe confifted of every fpecies that we had feen on the coaft of Kamtfhatka, and all the Aleutan iflands. I am inclined to think, that the birds, their eggs, and the fea animals caft on fhore, conftitute the chief food of the foxes in the fummer; and that early in the winter the ftraits freeze over, when they pafs to the oppofite ifland, which, from the verdant appearance of the low lands, feems likely to afford them edible roots for their fupport during a long winter. I did not obferve any fragments of fhells of any kind on the beach, nor the leaft traces of any inhabitants. This ifland is about fix miles from north to fouth; and, to judge from appearances, it is nearly fquare in its form.

The oppofite ifland is about 14 miles from north to fouth; the fhores everywhere broken and uneven, forming bays, bounded by projecting rugged cliffs, and detached pinnacle rocks.

Several white bears fwam round the fhip while we were at anchor, and three of them made many attempts to get up the fhip's fide; but at length they all fwam to the large ifland. Captain Cook did not obferve thefe ftraits, but thought the whole was one ifland.

At midnight got under weigh, and on the 15th, with a gentle fouth fouth-weft breeze, kept a northern courfe. Our foundings were now never more than 40 fathoms, having gradually decreafed to that from 100 miles north of Oonálafhka, where we had 80 fathoms.

fathoms. At noon we had hazy and foggy weather; latitude, by account, 61° 6', longitude 187° 9'.

On the 16th, wind and weather continuing, we saw no land, and our soundings had gradually decreased by noon to 26 fathom, fine sand. The last 24 hours run was 106 miles north-east 27° 7', making our latitude 62° 39, longitude 188° 54'. In the afternoon steered north one-half west. The evening being dark and misty, we hauled the wind, the ship's head being west south-west with a gentle south breeze, to keep clear of Clerke's Island. Our soundings at midnight were 24 fathom, sand and small stones.

July 17. Wind south-east by south, very foggy and misty weather. At two A. M. we stood to the north. Our soundings decreased to 15 fathoms; shingles; and many birds were flying about the ship. At six A. M. soundings 12, 11, 10 fathoms, when we suddenly got the bottom with six fathom line. We immediately hauled the wind, which freshened upon us, and stood south by east. In this direction we soon deepened our water to eight, nine, and ten, when it again shoaled to six fathom. We now considered ourselves as embayed in Clerke's Island, and kept working to windward; the soundings regularly decreasing on both tacks. At noon, our latitude was, by account. 63° 23', longitude 189° 29'. Continued making short boards till two P. M., when we brought up in six fathom, sand and stones; but, the anchor not holding, we dropped a second, and immediately after saw low land from east north-east one-half east, to west north-west, distant in the nearest place about three miles, which was immediately after hid again in the fog. The wind increased to a brisk gale, and the weather was misty and rainy all night.

July

July 18. Being very fqually, with fhowers and heavy fogs, we got a fpare anchor ready. At ten A. M. a hard gale coming on from eaft fouth-eaft, we kept occafionally paying out cable. About noon, the cable of the beft bower parted, and we dropped our fpare anchor in four fathoms, and payed out 15 fathoms cable, having only three fathoms at the fhip's ftern. As the gale continued, and no land was to be feen on account of the thick weather, we got a large fpare anchor out of the hold, and belayed a cable to it in cafe of need. The night was very dark, with a rough fea, and breakers juft aftern.

The 19th, at four A. M., the wind became more moderate, but the fog ftill continued. The fea being lefs agitated, we lowered our boats, and fifhed up our beft bower. At noon, in a moderate wind, we fent a mafter's mate towards fhore to found, and weighed the two fpare anchors. We obferved a current to the weft at three quarters of a knot. The mafter's mate reported, that he found three fathoms pretty clofe in with the fhore, but that the furf was violent. At fix P. M. the fog clearing up, we difcovered feveral mountains covered with fnow from weft fouthweft to north one-half eaft, and low land to eaft north-eaft one-half eaft.

Sunday the 20th, fouth-eaft by eaft, a moderate breeze, hazy and mifty. At noon the weather cleared up a little, and we faw lofty mountains covered with fnow fouth-weft by fouth one-half weft, and a peaked mountain, feemingly at a great diftance beyond the high land north-weft by north; our obferved latitude being 63° 26′ 34″. At two P. M. we faw two men walking along the low beach, who made a ftand oppofite the fhip, and, having fomething hoifted on a pole, waved it backwards and forwards.

wards. We immediately hoifted our flag. Mr. Bakoff was fent with the baidar on fhore; but the breakers were fo violent, that he could neither land nor get within hale; he therefore returned at fix P. M. We obferved the variation of the compafs 24° 16 30″ eaft. For the better trim of the fhip, we filled fix cafks with fea water.

On the 21ft we had little wind from the fouth-eaft, with rain at intervals. At noon Captain Billings and feveral gentlemen went on fhore on the low beach. At eight P. M. a gentle breeze fprung up from the north-weft, and the weather cleared amazingly, I went to the main-top-maft head, whence I could plainly fee the fpit of land, where the Captain went afhore, join to a mountain bearing weft by north, diftant 10 miles, trending due eaft about 17 miles, where it terminated, leaving a paffage into the lake which appeared behind it, and upon which I perceived a large boat rowing toward the mountains. The extremity of this fpit of land I computed at eight miles from the fhip, in a direction eaft north-eaft one-half eaft; and in the fame direction three miles farther is a projecting mountain, which conftitutes a part of the ifland, from which the land takes a circular fweep north-weft to the top of the lake, continuing the circle to fouth by weft.

We now took the following bearings: the fouth extremity of land, as far as we could trace a communication of mountains by low land, fouth-weft 50°, diftant about 12 miles. The body of a mountain, which appeared a detached ifland, fouth-weft 32° 30′, about 20 miles. Another mountain, feemingly detached from fouth-weft 8° to fouth 23° weft, about 16 miles. The promontory neareft the extremity of the fpit of land north-eaft 78°, from whence it took another circular fweep to fouth-eaft 75°,

where

where we perceived high land at about 10 leagues; but could not difcern whether it was connected, or formed a feparate ifland; and feveral intervening mountains were in the fame ftate of uncertainty. At nine P. M. Captain Billings returned on board, and we immediately got under weigh. He faid, the fea broke fo violently on the beach, that it was with great difficulty they effected a landing; and the Oonalafhkan, who had accompanied them in his fmall baidar, had had it dafhed to pieces. Obferving a foot-path on the fpit of land which was only 20 yards wide, he walked along it, in hope of meeting with fome of the natives at habitations which appeared at no very confiderable diftance. The fhore was almoft covered with the bones of fea animals. He paffed feveral dogs that were very tame; and, at the diftance of about three miles from the landing place, he faw feveral fcaffolds fix feet high, evidently for the purpofe of preferving, and keeping out of the reach of dogs, &c. fifh and fea animals; but no habitations were near. The failors near the boats obferved a very large baidar croffing the lake from the vallies on the oppofite fide, containing, as they fuppofed, about 30 men. Upon feeing this boat, one of the men walked along the path which Captain Billings had taken; but, not feeing him, and the boat advancing very faft, he fired his mufket, as a fignal for the Captain; upon which the boat immediately ftood back with all poffible fpeed. In confequence, they had no intercourfe with the natives.

Thurfday the 22d, by five A. M. having fteered fouth foutheaft, eaft, and eaft by north, we made 23 miles fouth-eaft 79°, when we rofe more land a-head, and were fully convinced that all the mountains between which there feemed to be ftraits, were joined by low land, The appearance, however, greatly juftifies

Lieuten-

Lieutenant Synd in placing fo many iflands in thefe parts. Our glaffes difcovered all the vallies occupied by the buildings of the natives, and fcaffolds for preparing or drying fifh and the flefh of fea animals.. Numbers of large boats alfo were hauled on the fhores; fo that this ifland muft be very populous. By noon we were off the fouth-eaft extremity of the ifland, at the diftance of one mile and a half; our obferved latitude 62° 55′, having failed from our place of anchorage 43 miles eaft fouth-eaft. We now doubled this cape, off which are two fmall iflands, the largeft about one mile in length, narrow, and replete with huts and fcaffolds; behind which we thought there was every appearance of good anchorage. The foundings were very uniform, according to our diftance from fhore; 12 fathom at the greater diftance, gradually decreafing, as we approached the land, to five and four fathoms at one mile and a half.

Having cleared this ifland, we ftood north and north-eaft about 20 miles, when the north-eaft extremity of the land bore weft, having a mountainous appearance, and terminating in a bluff-headed cape. [We did not fee Anderfon's Ifland.]

The wind fhifted to weft fouth-weft, and we fhaped our courfe north-weft by weft, with foggy weather. At noon our latitude, by account, was 63° 43′, longitude 192° 7′. The afternoon was hazy, with rain, and a gentle foutherly breeze. At eight P. M. the wind veered to north-eaft, and foon blew hard.

The 24th we had a brifk gale from north north-weft, on account of which we laid-to under main and mizen about three hours, when we ftood away eaft north-eaft to get well clear of the eaft of Clerke's Ifland, which would otherwife prove a lee-

I i

fhore.

shore, if the gale should continue; and the very narrow escape
that we had already experienced made us rather fearful of using
too much freedom with this island. At noon we got the sun's
altitude in the haze, latitude 64° 4′ 26″. Afternoon cloudy:
kept our course till midnight, with soundings at 19, 18, and 17
fathoms.

The 25th we had a hard gale north north-west with a rough
short sea, and laid-to under main and mizen till noon, when we
got the sun's altitude; latitude 63° 26′ 23″, longitude, by account,
193° 20′. We now steered north north-east under close-reefed
top-sails, the wind north-west, making two points and a half lee-
way, with 17 fathoms sandy bottom. At 3° 47′ 25″, apparent
time, the sun and moon's distance made our longitude 192° 24′
45″, latitude 63° 28′ 30.″ At eight P. M. we wore ship, steering west by south till the 26th, at seven A. M. when we saw
Clerke's Island right a-head. The wind blowing from the west
a gentle breeze, we let out all reefs, and set top-gallant sails. At
noon, our latitude, by account, was 63° 10′ 41″, hazy. In the
afternoon, with light airs, we kept a northern course. On Sunday the 27th, in the morning, we had foggy weather, with little
wind from the north-west. At eight A. M. it cleared up a little,
and at noon we observed the latitude 63° 31′ 8″, longitude 192°
55′. The afternoon being clear, with little wind, we kept a course
north-east by east all day. The 28th, a gentle breeze west by
north, steering north by west. At ten A. M. we saw high land
north-east 7°, and low land north-west 10°. At noon the latitude observed was 64° 12′ 19″, Sledge Island bearing north-west
6°, distant 12 miles. At one P. M. the continent of America
bore from north-west 55° to north-east 35′. At four P. M. being
about eight miles south of the nearest land, we cast anchor in 12

8

fathoms;

fathoms; our latitude being 64° 20′, longitude, corrected from our laft lunar obfervation, 164° eaft, Sledge Ifland fouth 78° weft, diftant nine miles, Cape Rodney north-weft 75°, alfo nine miles.

Captain Billings ordered the boats out, and went on fhore with the naturalift, draftfman, Mr. Bakoff, and Enfign Bakulin, with a few foldiers armed, befides failors in the long-boat and fkiff. We obferved a current fetting to the eaft, at half a knot.

In the morning of the 29th we had very light airs from the weft, with cloudy weather. At fix A. M. a baidar containing nine of the natives rowed alongfide, and came on board upon the firft invitation, leaving their arms in the boat, confifting of bows, and arrows pointed with green jade, calcedoni, and ivory; the bow ftrengthened, and rendered more elaftic, by the finews of the rein-deer, which were artfully bound round it. They had alfo lances about feven or eight feet long, fome pointed with iron, but very few; the generality being pointed with the tufks of the morzfh or walrofs very neatly cut. Upon one of them they hoifted a bladder, which fignal we anfwered with a flag, and they immediately came on board. They were well limbed, rather tall, had fine open and agreeable countenances, and were handfome and healthy. Their drefs was very neat and clean; being half-boots, neatly embroidered about the inftep with different coloured hair and finews, made of beautiful white leather, and tied round above the ancle with narrow flips of red leather, the foles made of bears' hide fmoked; tight, well made pantaloons, alfo of leather dyed yellow or red; a very neat park, refembling a carter's frock, reaching down to the knees, rounded before and behind, fo as to form two flaps, and open at the fides up to the hips. They wore no covering on the head;

and

and the hair was cut almoſt as ſhort as if ſhaven. They regarded every thing on board with admiration, but did not appear to be of a pilfering diſpoſition. I gave each of them a glaſs bead, and they immediately expreſſed great anxiety to obtain more, but had no articles of barter with them. They exchanged a few very neatly-made adzes of green jaſper, or jade, extremely ſharp ; nor did they ſcruple to part with their arms, and even ſtripped them-ſelves, giving their clothes for beads, knives, &c. One of them by accident broke a ſmall pane of glaſs, which threw them all into a ſtate of diſmay. He immediately offered his lance by way of indemnification ; but we made him underſtand that it was no loſs, by putting in another, and laughing at his concern, which pleaſed them all very much. They left us, making ſigns that they would ſoon return with ſome articles of trade for beads, &c. and paddled away towards Cape Rodney.

At ten P. M. the ſkiff returned with the Captain and Enſign. The long-boat, with the other gentlemen, did not get on board till the 30th towards midnight. The following account of their excurſion I tranſlated from the original remarks of one of the gentlemen in the party.

" We landed on a ſandy beach near the diſcharge of a ſmall
" river ; hauled both boats on ſhore ; and made a fire with drift-
" wood, which was in great plenty. The Captain, Mr. Bakoff,
" the Doctor, and Draftſman, walked along a narrow path on
" ſhore, quite unarmed ; and at a ſmall diſtance from the boats,
" we ſaw two natives coming toward us. When advanced with-
" in a few fathoms, they made a ſtand ; upon which beads were
" ſhewn them, and a few thrown on the ground. They were
" armed with lances, and advanced with the points toward us ;
                                                              " but

" but upon feeing the beads, and obferving our figns of friend-
" fhip, they turned the points of their lances behind them, and
" approached without hefitation.   Upon the firft fight of the
" natives, our interpreters were fent for, viz. the Oonalafhkan,
" the American taken from Kadiak, and an Anadyrfky Coffac,
" whofe mother was a Tfhutfki woman.   This latter they un-
" derftood perfectly well, and, embracing him upon his fpeaking
" the language of the Tfhutfki, we concluded that they were of
" that nation, and not Americans.   We returned all together to
" the boats, and Captain Billings gave each of them a copper me-
" dal and a few beads.   Shortly after we were joined by two other
" Americans, and obtained of them, for beads and a few uniform
" buttons, their bows, arrows, and lances.   Upon their invita-
" tion we accompanied them to their dwelling, leaving only four
" men armed to guard the boats.   The habitation was fituated
" four verfts from our landing-place ; and upon our arrival fkins
" of rein-deer and other animals were fpread for our feats before.
" the fire.   When we were placed, the hoftefs prefented each
" with a thin flip of the fkin of a marten, and immediately after
" with fifh, and the meat of the deer boiled ; but the intolerable
" ftench of the hut took away all appetite on our part.   It was
" dark when we arrived at the habitation ; fo that we knew not
" its extent, nor the number of its inhabitants ; notwithftanding
" the friendly behaviour of the natives, therefore, and though we
" were well guarded by our foldiers and failors, armed, and keep-
" ing a regular watch, we paffed a fleeplefs night.

" In the morning of the 29th we difcovered that we were not
" in a village, but in the temporary tent of a fingle family, pitch-
" ed for the fake of fifhing, and hunting wild-deer.   The tent
" was

" was covered with leather, except on one fide, which confifted
" of the inteftines of fea animals for the admiffion of light.

" At noon we returned to our boats, where, we were inform-
" ed, feveral natives had been, and traded with the men, giving
" them martens' fkins, the river-otter, and foxes, for beads; and
" that they invited them to their habitation, pointing out the di-
" rection, which we followed, along a fmall path of about five
" verfts. When we arrived there, we were alfo treated with the
" greateft friendfhip, and received in return for our prefents, the
" fkins of martens, foxes black and red, lynxes, and gluttons.
" Blue glafs beads, iron, and metal-buttons, were their favourite
" articles of barter.

" The Captain returned to the boats at feven P. M. where he
" found the Enfign; and, immediately embarking in his fkiff,
" with him and four failors, rowed on board. Dr. Merck was
" collecting plants and other natural curiofities; and the different
" hands were fcattered, fome trading with the natives; fo that
" it was near dark before we were all collected.

" Mr. Bakoff bought a baidar of the natives, in which he
" placed four failors; and, after taking fome refrefhment, we
" put off together. There was but little wind, and that was right
" in our teeth. The fea was rough, and the current againft us.
" We rowed about two hours, when the wind frefhened, and it
" rained hard. Having a fmall kedge on board, we brought up
" to wait day-light, very wet and much fatigued, and had loft
" fight of the baidar.

" At

Plate X

Fig 1

Fig 2

Fig 3

Fig 4

Fig 5

Fig 6

Fig 7

MISCELLANEOUS

Fig.1 A Pick-ax used by the Tshutzki made of the Tusk of the Morzsh. Fig 2 Stone Hatchet of
America, with its Case Fig.4 Sepulchre of Oonalashka. Fig.5.6 & 7. Baidar Dress and Bow used
by the Natives of both Continents at Berings Straits. Fig.3 an Instrument used by the Tshutski.

Published March 2.d 1802. by Cadell & Davies Strand.

" At day-break it cleared up a little, and we faw the fhip ;
" upon which we weighed, and took to the oars ; but, the wind
" frefhening with a head fea, we made but little way ; and after
" fix hours labour, the fea breaking into the boat frequently,
" which kept fome hands conftantly bailing out the water, all
" wet and exhaufted, we hoifted a fail, and ftood back for fhore.
" We ran on the fandy-beach near our former landing-place,
" cold, and almoft helplefs, with no means of making a fire ;
" but, to our inexpreffible joy, fome embers of the large fire
" which we had made of the drift-wood were ftill burning, and
" thefe enabled us to dry ourfelves. At four P. M., both wind
" and fea being much abated, and pretty clear, we again pufhed
" off, and reached the fhip by midnight, but heard nothing of
" the baidar, for the fafety of which we were under great appre-
" henfions *."

From our fears, however, we were relieved at four A. M. of
the 31ft, by her fafe arrival on board. The failors faid, that
rowing about in the dark and rain, without knowing where, they
were caft on fhore about 10 verfts to the weft of our landing-
place ; and that, notwithftanding the violence of the furf, and
the hollow waves, the baidar did not fhip a fingle fea. They
faid, that they were furrounded by the natives at day-light, and
traded with them ; but gave them a very bad character. I cannot guefs what articles of trade they had ; but they obtained feveral fkins of black and red foxes, martens, &c. I hope that the
natives had not the greater reafon to complain.

* Captain Billings told me, that he faw very neat earthen pots, in which the natives
dreffed their food, and that they had bowls and buckets of wood, with wooden fpoons ; that
he faw their armour, fome made of wood, and fome of bones, refembling thofe at Prince
William's Sound.

At

At eight A. M. we weighed anchor with a gentle breeze from the fouth, fhaping our courfe weft and weft north-weft; but, falling calm at two P. M., and getting into a current of one mile and a half weft, we brought up between Sledge ifland and the main. A large baidar full of natives, and two fmall ones, rowed alongfide; but before they came clofe they fang a fong, and made feveral antic motions. In token of friendfhip, they had a bladder hoifted on a pole; however, they would not be perfuaded to come on board, but exchanged feveral articles of curiofity for beads, &c. At eight P. M. we again got under weigh, with a gentle north-weft breeze, which foon fhifted to the weft and weft fouth-weft, with cloudy and hazy weather. At noon of the 1ft of Auguft, our latitude, by fhip's reckoning, was 64° 40'; longitude, corrected from our obfervation of the 25th July, 192° 27'. At four P. M. we faw King's Ifland, which is very lofty; the fummit broken and irregular; replete with pinnacle rocks; round in its appearance, and about five miles in circumference. We had a moderate fouth fouth-weft breeze, and our courfe was weftward.

On the 2d Auguft we ftood for the bay of St. Laurence. At nine A. M. faw the three iflands in the mid-channel of Bering's Straits. Our latitude at noon was, by obfervation, 65° 23' 50", corrected longitude 190° 37', when we faw the promontories on both continents, and the interjacent iflands. At eight P. M. we took the following bearings: a promontory on the continent of America, north-eaft 49°. Firft ifland, north-eaft 35°; fecond, north-eaft 18°; third, north-eaft 9°. The eaftern Afiatic pro-montory north-weft 29°.

We

We had variable light airs, and calms, with hazy weather, till Sunday the 3d, at fix A. M. when a gentle breeze fprang up from the fouth fouth-eaft, which made us ply to windward, making fhort boards for the bay of St. Laurence; till the 4th, at noon, when our latitude was 65° 37', longitude 189° 18'. The wind fhifting northerly, we ftood into the bay, and at four P: M. came to anchor.

CHAP.

CHAP. XVIII.

Reception by the Natives.—An Adventure of the Author.—Singular kind of Encampment.—Barter with the Male, and liberal accommodation with the Female Inhabitants.—Division of the Tʃhutʃki Nation into two Tribes ; which are separately deʃcribed.—Captain Billings, with a Party, leaves the Ship for the purpoʃe of a Land Excurʃion acroʃs the Country to the Kovima.—Tʃhutʃki Paʃtimes, &c.—Captain Saretʃheff, purʃuant to order, ʃails for Oonalaʃhka, and anchors in the Harbour of Illuluk.—Captain Hall arrives.— Preparations made for a Winter's Reʃidence.—Tribute collected.— Miʃʃion from Shelikoff's Eʃtabliʃhment at Kadiak.—Sketch of the Natural Hiʃtory of the Iʃland of Oonalaʃhka.—A Hurricane.

W E had no ʃooner dropped our anchor than a baidar full of Tʃhutʃki came along-ʃide, and ʃhewed us a paper from Kobeleff; who (they ʃaid) was now at the eaʃtern promontory ; and Dauerkin was with his relations toward Anadyrʃk ; but they added, that we muʃt come on ʃhore and give them ʃome tobacco before they would let us have the paper. Captain Billings went on ʃhore in his full uniform, and was received with every token of friendʃhip and reʃpect. Dauerkin was immediately ʃent for, and our intercourʃe with the natives was perfectly free and unguarded. On the 6th, at noon, Dauerkin arrived, with 12 large baidars full of Tʃhutʃki, whoʃe numbers increaʃed hourly. They had plenty of ʃkins of foxes, martins, hares, and the muʃk-rat of America, whence, indeed, they obtain the greateʃt number of their furs,

boats,

boats, and arms, in exchange for such articles as they get from Izſhiginſk, and from the wandering peddling traders about the eſtuary of the Kovima, &c.

On the 7th, I went on ſhore in uniform, but was not very well pleaſed with the reception that I met with. I had ſtrolled among the Tſhutſki to ſome diſtance from our tents and people, where one of the natives began to cut the buttons off my coat. I ſtruck him on the ſtomach with my fiſt, and he fell over ſome looſe ſtones behind him. One of our men (Vaſſiley Tolſtichen, a native of Anadyrſk), obſerving the tranſaction, ran towards me; the man got up and laughed, not ſeeming to be the leaſt offended at the blow. Tolſtichen told me, that they always inſulted little men, and ſuch as were leſs active than they. Upon hearing this, I challenged any one of them to run or leap. One of them offered to run with me to a point of land at leaſt a mile diſtant, and back again; this, however, I rejected, and propoſed running toward the boats, a little more than 200 yards. Arriving firſt at the goal, I received the pleaſing compliment of their acknowledging that I was, indeed, a man, though but a very little one. Not being inclined, however, to perform for their entertainment, I returned on board, fully reſolved not to quit the ſhip again ſo long as ſhe remained here.

The beach was now covered with the baidars of the natives, hauled on ſhore and turned keel upwards, one gunnel reſting on the ground, the other ſupported on their paddles: thus they ſerved the purpoſe of tents; and old dreſſed deer-ſkins ſewed together were uſed inſtead of curtains for the open ſide. Here the natives, men and women, ſlept indiſcriminately. The former traded with their dreſſes, furs, tuſks of the walroſs, whales' fins,

and

and pieces of the gut of rein-deer ſtuffed with chopped meat, marrow, and fat. The latter were extremely happy to grant any favours for beads, buttons, tobacco, &c. and that even in the preſence of the men, who actually introduced our people to the women when they had no other articles of trade. Theſe, however, were not their wives, but priſoners taken from their American neighbours, with whom they are frequently at war. The cauſe of the laſt affair between them was this: both parties meeting, on the chace of ſea animals, quarrelled; an engagement commenced, in which the Americans took one baidar and made the crew priſoners; the other, returning, procured a reinforcement, made a deſcent on the American coaſt, carried off a few women, and then peace was reſtored.

The Tſhutſki nation is divided into two very diſtinct tribes: the one is called Stationary, or fixed inhabitants of the coaſt; the other, Reindeer, or wanderers.

The former occupy ſuch places as are convenient for fiſhing and the chace of ſea animals, from the river Anadyr to a ſmall diſtance north of the eaſtern promontory. The extent of their population, according to the beſt intelligence that I could obtain, amounts to about 3000 males. Their chief habitations are about the bay of Anadirſk, particularly in the vicinity of Serdſi Kamen *, and in the gulph of Metchickma, which is between the

* Serdſi Kamen is a very remarkable mountain, ſituated in the north-eaſt part of the bay of Anadirſk, and projecting into it. The back or inland part is replete with cavities, whither the Tſhutſki fled when attacked by Pavlutſki. Here, ſecreting themſelves in the cavities of the rock, they ſhot great numbers of the Ruſſians on their paſſing by. Pavlutſki had at this time but a few of his followers with him, and returned to Anadirſk for a reinforcement; where he related, that the Tſhutſki ſhot his people from the heart of the rock; whence it acquired the name of Serdſi Kamen, or the heart-ſtone.

bays

bays of Anadirſk and St. Laurence. North of the eaſtern promontory the dwellings are but few, becauſe the ſea is not ſo prolific of fiſh, nor are there any foreſts; but the marine animals are more numerous, which is the cauſe of its being frequented on the chace; which ſometimes induces them to paſs the Shalatſkoi promontory into the Tſhaoon bay; which, they ſay, is about 15 days' journey from the eaſtern cape, ſleeping on ſhore every night. They were in this bay two ſeaſons waiting for our expedition from the river Kovima,—I ſuppoſe in 1787 and 1788.

They appear very induſtrious, and are neat workmen, which is evinced by their baidars, lances, arrows, bows, apparel, utenſils, &c. with which they ſupply the wanderers. They alſo trade with their female priſoners, receiving in return rein-deer, copper and iron kettles, knives, beads, and ſuch articles as the rovers obtain from the Ruſſian traders.

They dig cellars, in which they keep their ſupplies of food and oils. The proviſion conſiſts of dried meat of ſea animals and deer, roots, and berries. They regard the lips and ſnout of the morzſh, or walroſs, as a great delicacy when boiled almoſt to a jelly. The oil of the ſea animals they keep in ſeal-ſkins, and of this they obtain immenſe quantities; it not only being uſed for food, firing, and light, but alſo conſtituting a great article of commerce with the wandering tribe.

Kobeleff and Dauerkin have publiſhed very wonderful accounts of theſe people. Among other ſtories, they relate, that " the Tſhutſki, when aged or ill, require their friends to kill " them, which is immediately performed, as well with women " as men; and that a ſharp knife is the only remedy for all dif-
" orders."

" orders." But this they positively denied. I discovered by means of Tolstichen, that the aged were subject to rheumatic complaints, which they cured by lighting the dried leaves of worm-wood, so prepared as to burn like tinder, and letting it remain till burnt out on the affected parts: a custom also observed by the Yukagers, Tungoose, and Yakuti. That if they had any swellings from wounds, splinters, or any other cause, they applied a poultice, composed of chewn edible roots, moistened with fresh oil; and in cases of severe illness, offered sacrifices of deer to the spirits of torture; and sometimes a dog was killed, the sick led round it, and anointed with its blood and fat. In case of death, the body is burnt to ashes; stones are laid on the spot, to resemble in some degree the body of the man; a large stone at the head, anointed with marrow and fat; and the horns of deer form a pile or heap at a small distance. This place is visited once a-year by the relatives, who recapitulate the feats and actions of the deceased, by way of remembrance, when each of them adds a horn to the heap, and anoints the head stone.

I was not able to learn any particulars of their religious rites and ceremonies, nor any remarkable customs. They reckon only two seasons in the year, summer and winter; at the commencement of each of which they make sacrifices and merriments, in gratitude for what is past, and as an invocation for future success.

Kobeleff asserts, " that the wandering Tshutski make a practice " of lending their wives to strangers, as a mark of friendship; " and that they frequently exchange them amongst one another " for a short time." This, however, is not the case; for these people are extremely attached to their wives and progeny; and if one of them were inconstant to her husband, she would be

8

abandoned

abandoned by all : nor can a greater odium be thrown on a Tſhutſki woman, than to ſuſpect her guilty of favouring a ſtranger.

The wandering tribe conſider themſelves as a ſuperior race of beings, and the moſt independent of men. They call all the nations that ſurround them old women, only fit to guard their flocks, and be their attendants ; particularly the Koriaki. Reindeer are their only riches : theſe, and the ſkins of ſuch animals as they kill in their wanderings, they exchange with the Ruſſians, &c. for kettles, knives, and trinkets, which articles procure them arms, dreſſes, ſlaves, &c. from the ſtationary tribe. Their cuſtoms are alike, as is alſo their language.—This is all the intelligence that I could obtain of theſe people during my ſhort ſtay.

On the 12th Auguſt, Captain Billings, being completely ready to leave the ſhip, ſelected for his companions acroſs the country to the Kovima the following gentlemen :

Our naturaliſt,	-	Dr. Merck ;
His aſſiſtant,	-	Mr. Main ;
Maſter, or Sturman,	-	Mr. Batakoff ;
His mate,	-	Mr. Gileeff ;
Draftſman,	-	Mr. Varonin ;
Surgeon's mate,	-	Mr. Leman ;
Tranſlators,	-	{ Dauerkin and { Kobeleff ;

Attendants, two ſoldiers, and the Captain's cabin boy : in all twelve. (Kobeleff, not arriving here in time, was to join the company at an appointed place on the road, at no great diſtance.)

Captain

Captain Saretſheff received directions to ſail to Oonalaſhka, to collect tribute from the natives of all the neighbouring iſlands during the winter; and early in the ſpring to proceed to Kamtſhatka, where Captain Billings ſaid he would join us. Similar orders were to be left with the inhabitants of the bay for Captain Hall, to be given him upon his arrival.

In the evening the whole party took leave of the ſhip's company and went on ſhore, intending to ſet out on their journey early the next morning. The immenſe quantity of articles that Captain Billings took with him for preſents, to ſecure the friendſhip of the Tſhutſki, and enſure his own ſafety, appeared to me more likely to have a contrary effect. However, he ſeemed convinced that there was no danger to be feared from the natives.

On Wedneſday the 13th Auguſt, early in the morning, we ſaw the baidars of the Tſhutſki launched, and loading with the baggage of our friends; and at nine o'clock they departed in fifteen baidars, taking with them our moſt ſincere wiſhes for their proſperity and ſucceſs. The proſpect, indeed, was but a melancholy one.

There were now only two families of the Tſhutſki left in their tent, and Captain Saretſheff and I went on ſhore. We ſaw ſeveral boys ſkipping with a rope, and learnt that this was a favourite exerciſe, and very cuſtomary among the young women, of whom two held a rope, one at each end, and while they ſwung it round, a third ſtood in the middle to leap over it. We alſo obſerved boys and girls jumping on a ſkin in the ſame manner as we had ſeen them on the iſland of Tanaga; but the ſkin in this inſtance was that of a walroſs, with proper handles made of thongs

for

for fix or eight people to hold, which enabled them the better to catch and throw up the leaper. They alfo had a game of exercife refembling prifoners-bars, and threw ftones from a fling with great exactnefs.

I faw a woman dreffing a deer fkin with the hair on: it was, however, the latter part of the procefs; for it had been cleared of the flefh and filaments that adhere when taken off the animal, and had been covered with a coat of wet whitifh clay; which, being dry, fhe was fcraping off with a ftone fixed in a piece of ftick about two feet long, each end ferving for a handle: the ftone had a rough but not very fharp edge, and the fkin was faftened to a board. The whole procefs is exactly the fame as is practifed by all the Afiatic Tartars. For farther particulars concerning thefe people, I muft beg leave to refer my readers to fome remarks taken from the memoranda of Captain Billings's companions in his land excurfion through their country, and which will be found in a future chapter.

The Tfhutfki call Clerke's ifland E-oo-vogen; and fay, that it is the fame diftance from the north eaftern cape of the bay of Anadirfk, or Tfhukotfkoi Nofs; as is Kygmil (Cape Prince of Wales) from the eaftern promontory; that they pafs over in a day, and the ifland is extremely populous.

We made the diftance between the two continents 48 miles, the eaftern promontory bearing north-weft 42° from cape Prince of Wales, and the bay of St. Laurence from the fame point weft by north 62 miles, by true compafs. The three interjacent iflands are called, the firft Inalin, from the eaftern promontory 24 miles, bearing fouth-eaft 26°. Six miles farther, in a direction

eaft

eaft by north, is the fecond and largeft, Imaglin. Okivaki is the third and fmalleft, 10 miles diftant, fouth by eaft.

King's ifland they call Okiben, and Sledge ifland Ayak. The bay formed by the two capes, Prince of Wales and Rodney, is named Imagru, the deepeft part of which is the difcharge of a confiderable river called Ka-ooveren; near the fource of which, the natives fay, the country is well wooded. Kobeleff, fpeaking of a river in the vicinity of this place, relates, that on its border is a fmall town containing a church and oftrog, built and inhabited by Ruffians. He fuppofes them to be the remains of the fhipwrecked companions of Defhneff, a Ruffian adventurer who left the river Lena with feven veffels in 1648, and, having failed round the land of the Tfhutfki, arrived at Anadirfk alone, the other fix veffels being never afterwards heard of *. Notwithftanding all my endeavours, I could not find any body that knew aught of this matter, or had ever heard of any fuch place exifting.

At noon we returned on board, and immediately got under weigh. The Tfhutfki had promifed to give Captain Saretfheff fome frefh meat in the bay of Metfhikma; but he obferved, as we paffed this bay in the afternoon, that he could not weather the fouthern cape, if he entered with the prefent wind from the eaft of the north; he, therefore, thought it more prudent to purfue his voyage, than rifk the encountering any difficulties for the fake of a fmall quantity of rein-deer meat. He confidered the time alfo of too much confequence to be trifled with, the more efpecially as we had but a very bad fupply of fire-wood on board,

* For a particular account of this adventure, taken from original papers, fee Coxe's " Ruffian Difcoveries."

and

and were deftined to pafs the winter at Oonalafhka, which place produces none, except the ground willow, not exceeding the thicknefs of a walking ftick, and that only in a very few detached places between the mountains, difficult of accefs: a circumftance which made it abfolutely neceffary for us to endeavour to procure fome on our paffage, if poffible. We had feen a confiderable quantity drifted on the beach of Gore's Iflands, and this appeared the moft eligible place at which to procure it. The refolution, therefore, was taken, to direct our courfe for this place. The weather was very foggy; fo that we paffed the weft fide of Clerke's ifland, and to the eaft of the promontories that form the capes to the bay of Anadirfk, without feeing land. The wind continued from the north-eaft, and the weather remained wet and foggy; fo that we paffed Gore's Iflands without getting a fight of them. Nor did we dare venture an attempt to approach too near. Confidering our prefent fituation, nothing but Captain Saretfheff's anxiety about Captain Hall and the gentlemen left at Oonalafhka, prevented his fteering direct for the harbour of St. Peter and St. Paul in Kamtfhatka. This alone induced him to take the refolution of fteering direct for the ifland of Oonalafhka, which was the firft land that we made, and this we reached on the 29th Auguft, when we anchored in our old place in the harbour of Illuluk.

We were now informed, that Captain Hall arrived here a few days after our departure, and, having taken on board the gentlemen, ftores, &c. left for him, had followed us to the bay of St. Laurence. We were, therefore, in hourly expectation of his return; and on the 31ft he appeared in the offings, and the next day (the 1ft of September) came to anchor along-fide. We had hauled our fhip into a little cove behind a fmall rocky ifland. It

had

had the appearance of a pond 25 fathom wide, and 150 fathom in length; the north-east bounds were low land, but in every other direction lofty mountains. Four anchors were carried afhore, two on the ftarboard, and two on the larboard fide, from the head and ftern; and two cables were belayed to each anchor. The fmall veffel, which Captain Hall had named the Black Eagle, was moored along-fide, at the diftance of 20 feet. Thefe precautions were taken in confequence of the natives telling us, that the fqualls in winter were very violent, particularly in this part of the ifland. Here we laid up our veffels for the winter, and built a fhed on fhore, of yards, top-mafts, &c. covered with fails, in which we kept the articles and provifions that were landed. We alfo erected a working fhed of fods, thatched with rufh grafs, for the fail-maker, carpenter, block-maker, &c. The roof was formed of fpare yards, fpars, and oars; and the window-frames were conftructed of old cafk-ftaves. This building alfo contained two ovens for baking bread, &c. The commiffary, or rather purfer, Mr. Erling, a Ruffian, and I, built a fmall hut nearly in the fame manner; the infide of which we lined with whales' fins. The captains of both veffels, and other officers, retained their births on board; as did alfo the greater part of the crew; for the habitations of the natives were fo fmall, and formed fo entirely without conveniencies, that none of our failors or petty officers chofe to take a birth among them. Their neareft village was that of Illuluk, about a mile from the veffels; and that confifted of four or five huts, pretty deep under ground; the tops of which were overgrown with grafs and weeds, and prefented an appearance refembling heaps of earth: the entrance was at the top, through a fmall fquare hole, which alfo ferved for the admiffion of light, and the difcharge of fmoke. The want of fire-wood and other timber precluded the poffibility of erecting barracks; and the account

which

which the natives gave of the mildnefs of the climate juftified the choice of the veffels, which were the more convenient habitations. Befides, it fecured to all hands the rations of provifion and brandy, which are allowed only at fea or in a foreign port.

We had conftantly fome party or parties out in our boats, collecting the drift-wood on the beach, which, however, was in general fo fodden with fea water, that it would not burn; and they were fortunate if during the day they collected a day's fupply. Thefe parties were furnifhed with mufkets and ammunition for fhooting wild fowl, which were alfo fcarce and fhy. Numbers of hands were alfo fent inland to cut the ground-willows.

The natives having been informed, before we departed for the land of the Tfhutfki, that we fhould return to pafs the winter on this ifland, had caught and dried a quantity of halibut, cod-fifh, and falmon. They had alfo collected a confiderable quantity of berries in cafks which were left with them for that purpofe; and every poffible precaution had been taken to fecure frefh provifion for the winter; for our falted beef no longer poffeffed its nutritive juices, and our dried bread was almoft exhaufted: but we had with us a confiderable fupply of flour; fo that we only wanted fire-wood to bake it.

We now formed among ourfelves a little republic, in perfect congeniality of fentiment, complete friendfhip and harmony; equal in our manners and way of living; uncontrolled by feverity, yet obferving ftrict order and fubordination. I may fay, that the poffeffions, even the purfes, of each were fubfervient to the wants of the other. Our fociety confifted of, Captains Robert Hall, Gabriel Saretfheff, and Chriftian Bering; our furgeon-major

major Robeck; furgeon Allegretti; Meffrs. Bakoff, Bakulin, Erling, Pribuiloff, and myfelf.

Having thus defcribed our fituation and arrangements, I fhall
proceed to our occupations. The natives were informed, that
our orders were, to collect tribute, and to receive fuch as they
voluntarily chofe to give as an acknowledgment of their fubjection to the Court of Ruffia; but that we were not authorifed to
exact any thing beyond what they could conveniently fpare; at
the fame time they were defired to bring the receipts for the tribute which they had given to the hunters, or Ruffian Promyfh-
lenicks. Several of the inhabitants immediately brought black
and red fox-fkins, and received prefents for the fame, more in
their eftimation than equal to the value. In their fifhing parties
they fell in with the natives of other iflands, and communicated
the intelligence; fo that this part of our bufinefs was known to
all the natives of the weftern iflands, and to thofe eaftward as far
as to Kadiak, from which ifland Shelikoff's eftablifhment difpatched to us two of their companions, who were efcorted by
numbers of the natives of the different Aleutan iflands and of
Kadiak*. The object of their miffion was, to requeft a fupply of
medicines (with directions how to ufe them) for the venereal difeafe, which had arrived in their different fettlements at an alarming
height. They alfo were in want of many common neceffaries, as
tobacco, brandy, &c.; of the latter articles we could not fend
them any, Captain Billings having left us but a very fcanty ftock;
but of medicines our furgeon-major fent as much as he could

---

* I took this opportunity to prove the correctnefs of my Vocabulary, and to make particular inquiries about Alakfa; which they affured me was not an ifland, and that I muft
have been miftaken in fuppofing they called it a Kichtack; that no ftraits exifted; but
that they frequently carried their boats acrofs a narrow neck of land, and went down a
river to the north fide of this point of land.

possibly

possibly spare, with proper directions for using them. Numbers of the natives of the Aleutan * Islands, who accompanied the mission, complained of the treatment they met with, and wished to return to their homes: to the best of my recollection, they were liberated; the hunters, however, were told, that they must be answerable for the tribute of such as they kept in their employ, as also for the manner in which they treated them.

Shortly after our arrival at this island, several of our hands were afflicted with the scurvy, but in a flight degree; and such as chose to reside on shore were allowed a birth in the working-shop. We had malt, hops, and a considerable quantity of essence of spruce; and beer was brewed for the benefit of all hands, especially the sick. Berries were also administered, and every antiscorbutic that we could procure; but we did not perceive that any good arose from it.

The shocking accounts that we had heard of the ravages which the scurvy had made among the different hunters who had passed the winter on this island, and particularly the crew of Levasheff's vessel, who commanded the second ship in Captain Krenitsin's expedition in the year 1768 †, made every one of us dread the effects of this fatal disease; and, thinking the best way to guard against it was, to copy the natives in their mode of living, I made the chief part of my diet consist of raw fish, muscles, and limpets; using, instead of tea in the morning, a tea-

* By the Aleutan Islands, I mean the whole chain from the point of Alaksa westward to Kamtshatka, except Bering's and Copper Islands.

† This officer lost almost all his hands by this dreadful disorder; nor could he ever have left the island without assistance of men from Krenitsin's vessel, who had passed the winter in the straits of Alaksa.

spoonful

spoonful of essence of spruce in a small tea-kettle full of boiling water; and in the evening, we boiled beer with berries, sugar, and pepper, which, with the addition of some corn-brandy, was our substitute for punch.

I also daily collected a sufficient quantity of wild cresses to afford a sallad for our mess; and on Sundays procured enough for the whole company in the cabin. Two or three times a week I obtained also fresh fish, by frequenting the rocks at low water, which were overflowed at flood; and these I caught by the following stratagem: I baited a fish-hook with a raw muscle, and thrust it into the holes, or rather cavities, in the rocks: the fish lurking under these stones-took my bait, and I by these means sometimes in the course of half an hour caught half a dozen fish: the sorts were—the wrasse, the father lasher, a large species of the blenny, and the turpug. [For a particular description of some of these fish, I refer my reader to the APPENDIX.] The other fish are halibut, cod, two or three species of salmon, and sometimes, but very rarely, the tshavitsha, a species of salmon very common in Kamtshatka, about Neizshni, between four and five feet long; also thornbacks and flounders. The shell-fish are—different species of crabs, the small pearl oyster, muscles, cockles of an immense size, wilkes, periwinkles, a great variety of edible limpets, and the cuttle fish.

The birds that I observed were—two species of geese; one termed by the Russian hunters laidenoi. These appeared on the 31st August, two days after our arrival, and wintered here. The head, neck, and breast, are white; it has a large black spot on the throat; back, wings, and tail, ash colour; the extremity of the feathers barred with a black streak edged with white; bill

14

and

and legs yellow; claws black. They remained here until the 18th April, and on the 19th the other species arrived, which I have described at Kadiak, where we obtained some: these depart about the 30th of August. Toward the latter end of September a few of the snow-buntings came, but only stayed a few days.

The safka, a kind of duck very frequent in Kamtshatka, made its appearance at the beginning of October, and wintered. The 12th November we saw the turpan of Ochotsk; but these stayed only a few days, and were in very poor condition. There is also a species of bunting with a red head and breast; but of these I saw only two or three; they are much sought after by the natives, who ornament the strings to their darts and dresses with the red feathers: also a bird as small as a wren, which emits a delightful note: these and the safka, indeed, are the only melodious birds on the island. Here are also a few partridges, teals (tshirok), cormorants (urili), awks (ari), sea parrots (toporki), and gulls (tshaiki). A very large species of the gull kind was killed by a party collecting drift-wood in the beginning of April. They had retired to a cave to refresh themselves, when this bird, pursued by an eagle, took refuge among them. The wing had three joints, one more than I ever saw in any other bird. The Russians call them Semi Sazshenoi (seven fathom), from the extreme length of their expanded wings. They are frequently seen, but the natives have never found their nests or eggs. When picked, it appeared very like a large turkey, and, to our depraved tastes, was not inferior in quality when dressed. Eagles are numerous, as are also the glupysh, which I take to be Pennant's foolish guillemot.

M m   The

The only animals that I saw were foxes and mice; the latter, I obſerved, generally choſe the ſouthern ſide of the mountains for their burrowing places, and the freſh ground thrown up by them were the ſpots from which I collected the wild creſſes.

The morning of the 1ſt of April 1792 being clear, I roved about the ſouth ſide of the mountains to enjoy the ſun, which we had not ſeen ten times ſince our arrival on this iſland. During my walk, I ſaw, at the entrance to one of the mouſeholes, a conſiderable quantity of edible roots: theſe conſiſted of makarſhine, ſarana, and another root unknown to me, about the ſize of a coffee bean (but few of them): the quantity might be about ten pounds weight, thus brought into the ſun to dry by the mice, more provident than the human part of the inhabitants of this iſland. I alſo noticed, for the firſt time, that the ſweet plant of Kamtſhatka, the kutagernik, or wild angelica, the broad-leafed ſorrel, and kiprey, were breaking through the earth. The other productions of the iſland are, the ground willow, already deſcribed (but not a ſingle tree of any denomination whatever, nor does any of the iſlands weſt of Kadiak produce a tree of any kind: this I can poſitively aſſert); two berry-bearing buſhes, the tſhernika and golubnika, about eighteen inches high, on the ſouthern ſide of the mountains, and in ſuch places as are ſheltered from the north winds; the mountains alſo produce the ſhikſhu, or ſiecha, and wortle-berry. The vallies yield raſpberries, white, large, and of a watery taſte. The edible roots are, ſarana, makarſhina, and the root of the lupin; this plant bears a more beautiful flower than in Europe: the kutagernik is ſometimes uſed for food, mixed with fiſh ſpawn, I believe on account of its bitter flavour. Wild muſtard was plenty about the old habitations.

tions. The grafs is coarfe and rufhy; I am inclined, however, to think its quality fucculent; for it appears to me of the fame kind as grows about the harbour of St. Peter and St. Paul in Kamtfhatka, of which the cattle are very fond, and it fattens them extremely. The foil is not deep, but black and fine, unmixed with clay or loam. It was with great difficulty that we procured, near the fource of a rivulet, a fufficient quantity of clay to ufe as cement to our ovens, built with the ftones collected on the fea-fhore. Here are no rivers; but feveral rivulets, or fmall rills of water, run into the fea. There are two extinguifhed volcanoes on this ifland; and near one of thefe there was formerly a hot fpring, but it is now buried under ftones fallen from the mountain, which produces abundance of native fulphur. Earthquakes are frequent, and, by the account of the natives, fometimes very violent.

The fea produces, befide the fifh already mentioned, whales, grampuffes (kofatki), porpoifes (fwinki), the fea lion (fivutfha), and the urfine feal (kotic); the two latter ufed to pafs this ifland in great herds late in the autumn; but they have not appeared the two laft years, which I attribute to the havock made among them by the hunters on the iflands difcovered by Pribuiloff to the north of Oonalafhka. Sea otters are almoft forgotten here; but they fometimes appear on the rocky iflands off Atcha.

I fhall now return to our fociety. We had but little to do during the prefent year. Our foraging parties met with very ill fuccefs, although they were of material affiftance with the little fire wood that they obtained. They could not fhoot any game, which I afcribe to their being too numerous and noify: for I was fuccefsful when I went out alone, but found the wild fowl exceedingly fhy. We experienced a conftant fucceffion of mifts

and

and fogs; fometimes during the night the ftars appeared; we had frequent gales of wind, and very ftrong, and encountered one hurricane, which, probably owing to the furrounding lofty mountains, acted like a whirlwind upon our veffels, carried the Black Eagle on fhore, and, catching the Slava Ruffia, all her cables parted like pack-thread at one inftant; but, notwithftanding fhe was at the mercy of the gale, and in great motion in the eddy of the wind, its oppofite currents only drove her a fhort diftance along the bafin, and back again. We expected her every moment on the rocks; the violence of the hurricane, however, abated, and we again got her to the old moorings, without having received any damage. Several of our men were laid up with the fcurvy towards the end of the year, and we buried one young man, whofe death was occafioned by this diforder; he had re-fided on fhore from the time of our arrival.

CHAP.

CHAP. XIX.

Deplorable state of the Party under the Effects of the Scurvy.—Attention and Services of the Natives.—Sketch of the Religious Notions, Government, Arts, Manners, &c. of the Natives of the Aleutan Islands.—Mode in which the Russian Hunters carry on their Concerns.—Quit the Island, and arrive at St. Peter and St. Paul.—Find there the Alcyon, Captain Barkley, from Bengal, who, though having on board Articles of the first Necessity, which he offered at a very low Price, was forced to return without having disposed of any.—Reflections of the Author on Ventures of this kind.—Better Success of a former Adventurer, Captain William Peters, in 1786.

THE year 1792 had commenced with the most inauspicious prospects that the imagination can conceive: except myself, I believe, every one of our Company was affected more or less with the scurvy; some without any other outward marks than a sallow complexion, accompanied with shortness of breath, and an extreme lassitude of the whole frame, which prevented their taking ordinary exercise, or even walking far; some had small livid pimples all over the body, particularly about the legs, with soreness and violent itching; some had large livid blotches on their legs, arms, and other parts of the body; some were bloated all over, and almost all had their gums swollen to such a degree, that they nearly hid their teeth. Thus situated, it was with difficulty that we could muster able hands enough to hoist a cask of water on board.

13

The

The natives exerted themfelves to procure fifh; and, when the weather prevented their aquatic excurfions, they collected bundles of willows for firing. The bufinefs allotted to me was, the receiving of tribute from the Aleutans, diftributing prefents, and giving returns for whatever they brought us. When not employed about this bufinefs, I was ftrolling over the mountains gathering creffes, or at low water along the fea-fhore, fifhing among the rocks, or getting limpets, mufcles, &c. The birds were fo fhy, that I could but feldom fucceed in fhooting any. I was always alone; for, though feveral gentlemen frequently attempted to walk with me, they very foon became tired, and returned, leaving me to purfue my folitary perambulation.

The profpect before us grew more melancholy as the feafon for our departure advanced. More than three-fourths of our Company were confined to their hammocks by the fcurvy; but our Surgeon, Allegretti, was among the number of thofe who, with the affiftance of crutches, were enabled to move from place to place; and he, with Mr. Bakulin, was taken into our hut. Notwithftanding every poffible method was adopted by our furgeons to check this diforder, it raged with unabated violence; and, toward the latter end of the month of February, we fometimes buried three men in one day; and the moft athletic in appearance were the earlieft victims. It was equally deftructive to thofe who dwelt on fhore, as to thofe whofe birth was on board fhip. At this time we began to doubt the poffibility of ever leaving the ifland. I ftill continued my walks, but found fome difficulty in rambling over the mountains: it certainly fatigued me more than ufual; but I did not on that account fhorten my excurfions.

Early

Early in the month of March, the wind, which had hitherto blown from the northern quarters, veered to the fouthward; although rainy and mifty weather continued during the day, the fogs were lefs, and the nights more clear. We now obferved, to our inexpreffible joy, that the mortality ceafed; that thofe who were violently afflicted with the fcurvy did not get worfe, and that no more were laid up; and, foon after, appearances indicated returning health. The inhabitants of the ifland, with the natives who brought us their tribute, alfo fupplied us with abundance of halibut and cod; the wild muftard growing about the habitations was gathered, and diftributed to the different meffes; and we again revived, in hopes of better days than this ifland could afford.

During the winter, I had frequent opportunities of reading my vocabulary (taken in 1790 in the ifland Sithanak) to the natives, who underftood every word; and, therefore, I think I may venture to pronounce it pretty correct: on all the Aleutan iflands the *tb* is pronounced exactly as in England.

Of their religion I am not able to give fo particular an account as I could wifh, owing to their extreme fuperftition; for they believe, that the kugahs, or demons, of the Ruffians are more powerful than theirs; and that, ever fince thefe vifitors came among them, they have been fubjected to the greateft flavery and diftrefs; that if they have only mentioned their real name, it has been a fufficient means to lead to their difcovery and torment. " Some of us have even adopted their method of worfhip*, in " hopes of foliciting the protection of their kugahs, but without

* I have here taken the very words of the Aleute interpreter Elifey, who was chriftened; to which ceremony he alludes.

          " producing

" producing any falutary effects." The natives imagine, that the curiofity which their vifitors expreffed of feeing every cere-mony was merely with a view of infulting their kugahs, and in-ducing them to withdraw their protection ; by which means they fuppofe them to have fucceeded in compelling fubmiffion. The want of employment for their minds increafes their fuperftition, and they carefully avoid difcovering their magicians, or explain-ing any meaning in their ceremonies. They ftill obferve their annual dance in mafks, and with painted faces ; the mafks are called kugahs ; and I difcovered, that fome particular ornaments of their drefs ufed upon this occafion were regarded as charms, having power to prevent any fatal accidents, either in the chafe or in their wars ; but in the latter they now never engage. We were informed that the greater part of the inhabitants of Sitha-nak had been victims to illnefs fhortly after we left the ifland in 1790.

According to the beft intelligence that I could obtain of the population of all the Aleutan iflands, the number of males (in-cluding children) does not exceed eleven hundred, of which num-ber about five hundred of the moft active are employed by dif-ferent parties of Ruffian promyfhleniki, or hunters. Formerly, one village on this ifland contained more than the above num-ber. At that time they had one chief prefiding over the whole ifland, whom they called Kikagadogok, being chofen by the na-tives from among the Tokok * or Dogok, chiefs of villages (for it is pronounced both ways at different iflands). The reft are vaffals, diftinguifhed by the name of Talha. They fay, that

---

* I have obferved, that the chiefs of villages are called by fome of the Ruffians Tookoo; by others Toyon. Whence they obtained the name of Tookoo I know not ; but Toyon, or Toigon, is the proper Yakut name for Chief.

mankind

Plate XL.

Page 272.

Drawn by N. Alexander.

Engrav'd by W. Cooke.

Masks used by the Oonalashkans in their Dances, with the Darts used by the same people, and the two sides of the board from which they are thrown.

mankind were the offspring of dogs by the command of Aghu-guk; and that all of them came from the weft, where they fup-pofe there exifts an immenfe and very populous country.

Although they formerly had places wherein to depofit the pro-duce of the chafe, they never were accuftomed to lay in a ftock for the winter; for they only preferved their food until it came to their turn to feaft. As, however, at that time, the iflands were populous, and their villages extenfive, this method was nearly the fame thing; for the different villages vifited each other in re-gular rotation, and were guefts until the whole quantity that they had collected was exhaufted, which was not till their fifhing fea-fon re-commenced, when their magicians and the kikaga-dogok were confulted, and commenced their incantations for new fuc-cefs in the enfuing feafon, affuring their kugahs that nothing had been wafted of their former bounty.

They fifh with bone hooks. The lines are either a remarkable fpecies of fea-weed * feventy fathom long, or the fins of whales cut very thin and even †. Their darts for animals are coloured, fome red and others black; for they have different coloured paints, or earths, which they mix with oils of fifh; as white, blue, red, and black. Thefe they obtain from a mountain near the village Ama-da; but where that is fituated I know not.

The capacity of the natives of thefe iflands infinitely furpaffes every idea that I had formed of the abilities of favages. The or-

* A fpecimen of which is now in the poffeffion of Doctor Rogers, with feveral other Aleutan curiofities.

† The natives, when fifhing for halibut in 70 or 80 fathom water, frequently haul up with the line beautiful white fticks and their roots. Thefe are from fix to eight feet long, very thin, and without bark or branch. When firft taken out of the water they are as elaftic as whalebone; but, when kept a confiderable length of time, they refemble white coral, and are brittle.

N n   der

der eftablifhed among them, and their fubordination to fuch chiefs
as they have felected for their rulers, certainly originate from
principles of adoration which they poffefs for an exifting invifible
Superiority, and govern their conduct with that propriety which
feems moft likely to attain fecurity and protection, both in this
world and in the next; for they firmly believe in another world,
and imagine that fuch as live in conformity to the will of Aghu-
guk will there obtain all neceffaries with little trouble, and not
be under the control of the kugah. Their behaviour, therefore,
is not rude and barbarous, but mild, polite, and hofpitable. At
the fame time, the beauty, proportion, and art with which they
make their boats, inftruments, and apparel, evince that they by no
means deferve to be termed ftupid; an epithet fo liberally be-
ftowed upon thofe whom Europeans call favages. It is much to
be lamented, that they are under the fway of the roving hunters, who are infinitely more favage than any tribes that I have
hitherto met with; nor do I fee any means of checking their
outrages; for the authority of government can never reach thefe
diftant regions: the only profpect of relief appears to me to confift in the total extirpation of the animals of the chafe; and I
think I may venture to fay, from the daily havock made among
them, that a very few years will ferve to complete this bufinefs.

As I have fo frequently mentioned the hunters, a fuccinct account of their proceedings may perhaps not prove uninterefting
to my readers.

Their galliots are conftructed at Ochotfk, or at Neizfhni Kamtfhatka; and government, with a view of encouraging trade, have
ordered the commandants of thofe places to afford as much affiftance as they can to the adventurers; befide which, the materials
of

of the very frequently wrecked tranfport veffels, though loft to government, are found the chief means of fitting out fuch an enterprife, and greatly leffen the expence. The failors agree to the diftribution of fo many pais (fhares) among them, in lieu of wages : thus their veffels are procured and manned. The cargo confifts of about five hundred weight of tobacco ; one hundred weight of glafs beads; perhaps a dozen fpare hatchets, and a few fuperfluous knives of very bad quality ; an immenfe number of kleptfi (traps for foxes), and a fmall ftock of provifion, confifting of a few hams, a little rancid butter, a few bags of rye and wheat flour for holidays (for they do not make a practice of eating bread every day), and a confiderable quantity of dried and falted falmon. They are alfo fupplied with a few rifle-barreled guns, and a quantity of ammunition, for their defence againft the natives.

Being thus equipped, with (" Bozfhe Pomotfh ") God's help they go to fea. Upon their arrival at any of the inhabited Aleutan iflands, they formerly ufed to take a number of women and a few men as hoftages ; but now they take poffeffion of the village, and, after hauling their veffel on fhore, diftribute their kleptfi to the natives to catch foxes, and fend out parties to collect firing, to fifh, and to chafe fea animals. Some of the hunters go to the contiguous iflands, and exact the fame obedience from all, while they themfelves live in indolence and eafe. The articles of trade, as they call them, are given in fmall quantities to the women, to fecure their attachment; and the men are fometimes rewarded for a hard day's work with a leaf of tobacco.

Ever fince Shelikoff formed his eftablifhment at Kadiak, no other companies have dared venture to the eaft of Shumagin's

N n 2   ifland.

island. I am inclined to think that Suchanin's vessel will be the last that will attempt to visit these islands for furs; and probably he will obtain hardly any other than foxes', which are still here in considerable numbers, and even resort to the villages in cold nights in quest of prey.

Shelikoff has formed a project to obtain the sole privilege of carrying on this trade without a rival; and he will probably, one day or other, succeed; but not before the scarcity of furs lessens the value of this trade, and renders a fresh capital necessary for making new excursions to discover other sources of commerce, or rather of wealth; then the directors of the present concern will explore the regions of America; and, if nothing advantageous occurs, they will, doubtless, retire from the concern, secure in their possessions, and leave the new members to pursue the undertaking.

During the month of March the scurvy seemed perfectly at a stand, neither increasing nor diminishing materially upon the afflicted; but early in April, when the new plants produced a supply of vegetables, those sick who had used crutches were enabled to relinquish them, and willingly gave them up to such as began to creep out of their hammocks. The weather, though it continued hazy, was drier than it had been ever since our arrival.

We now began making preparations to leave this fatal island; when we discovered that our sails, cordage, and rigging of every kind, had suffered from the climate as much as our ship's company; every thing was quite rotten, and our vessels very foul. Captain Hall, who had now the command, took charge of the Slava Rossie, and Captain Saretsheff went to the Black Eagle.

Notwith-

Notwithſtanding every individual exerted himſelf to the utmoſt of his abilities, however, we were not ready to depart before the middle of May.

We had received, as tribute from about five hundred of the natives of the Aleutan iſlands, a dozen ſea otters' ſkins, and of fox ſkins, I believe, near ſix hundred of different ſorts; in return for which, we had diſtributed all our trinkets and tobacco. The extreme poverty of this place prevented our obtaining any articles of value for ourſelves: we procured, indeed, a few curioſities, but nothing elſe.

On the 16th of May our veſſels were hauled into the outer bay. We were now elated at the proſpect of once more reviſiting Kamtſhatka, after the melancholy ſenſations that we had endured for eight months and ſixteen days, paſſed in one continual ſtate of anxiety upon this iſland, the grave of ſeventeen of our ſtouteſt hands; where, during the whole of our ſtay, we had only been cheered eighteen times with the ſight of the ſun, and never experienced one clear day. On the 17th we ſailed out of the Bay of Amoknak, and the ſame day ſaw the very remarkable ſolitary rock, reſembling a pillar, ſituated about 30 miles north of the eaſtern point of Oomnak.

Nothing remarkable happened during our paſſage to Kamtſhatka. We loſt ſight of the Black Eagle the 7th of June; and on the ſame day ſaw an iſland, which we took for Semi Sopoſhni, burning in ſeveral places, particularly toward the ſouthern extremity. On the 16th, after encountering a few contrary gales and baffling calms, we arrived in the bay of Avatſha, in a very thick fog (which fell upon us at the mouth of the bay), and came

to

to anchor near the entrance into the inner harbour of St. Peter and St. Paul, without being able to fee any land *.

Notwithftanding we were as filent as poffible on board, with a view of furprifing the inhabitants when the weather became a little clear, we had not lain long before we heard a boat rowing towards the veffel; and were fhortly after amazed at feeing an Englifh pinnace coming along-fide; with Captain Charles William Barkley in it, whofe veffel, the Alcyon, from Bengal, was at anchor in the inner harbour on a trading voyage. His cargo confifted of articles that were invaluable in this part of the world; particularly in a port fo eligibly fituated for encouraging commercial undertakings; namely, iron in bars, anchors, cables, and cordage, with various kinds of ironmongery wares, and a confiderable ftock of rum. Notwithftanding this, the commander of the port having neither authority nor refolution to fecure a purchafe for account of government; and the traders of this peninfula (who ftile themfelves merchants) being merely a fet of roving pedlars, without either capital or credit (and, what is ftill worfe, without principles to fecure either); Captain Barkley was neceffitated to take thefe articles back again, although they were offered at lefs than one third of the charges of tranfporting fuch commodities from the manufactories in Siberia.

A man who has refolution to ftrike out a new line of commerce, or rather to feek a new fource of trade, in parts of the world fo little known as are thefe regions, at the fame time unacquainted with the language and with the wants of the inhabitants, is rather threatened with lofs, than flattered with profpects of profit,

* Captain Saretfheff, in the Black Eagle, arrived on the 19th.

in

in the firſt attempt; and nothing ſhort of enthuſiaſtic hope of future advantages can compenſate for the degree of anxiety that he
muſt ſuffer.   Such a man, moſt certainly, merits all the encouragement that the government can give him, which is ſure to be
eventually benefited by his ſucceſs.   Conſidering theſe circum-
ſtances, and that the two veſſels employed in our expedition were
in the greateſt need of entire new rigging, anchors, &c. the preſent favourable opportunity of ſerving Captain Barkley by clearing his ſhip was a ſecondary conſideration, compared to the advantages which government would have derived from ſo valuable
an acquiſition of the moſt neceſſary articles that the port could
poſſeſs.   This I repreſented to the governor of the port, and to
the commanding officers of our expedition; but both equally
feared to act without poſitive orders.   In other reſpects, however,
we gave him all the aſſiſtance in our power.   Captain Barkley
was accompanied by his lady, and a ſon of about ſeven years old.
Their behaviour was very polite, and particularly pleaſing to us.
I lament that we were not able to make them equal returns, but
flatter myſelf that they were ſatisfied with our endeavours.   The
extreme poverty of the place, and the miſerable ſituation that we
were in, muſt have been ſufficient in their eyes to prove an excuſe
for us.   They left this place the 1ſt July O. S.

Captain William Peters, who arrived here on the 9th Auguſt
1786, was more fortunate in the diſpoſal of his cargo, owing to
a mere accident that befel the only man in this part of the world
who had a capital and an eſtabliſhed credit in Moſcow, which ca-
pacitated him to become a purchaſer: I mean Gregory Shelikoff,
who ſailed the 22d May of the ſame year from his eſtabliſhment
in Kadiak for the port of Ochotſk, with a cargo of furs.   Contrary
winds prevented his arrival at the Kuril iſlands till the 30th July,
which

which also detained him here eight days. Still continuing westerly, he resolved to steer for the bay of Tshekafkoi, at the estuary of the Bolshoia Reka, on the west side of Kamtshatka, to purchase a supply of fresh fish. When he arrived off this place he cast anchor, and went on shore with the ship's boat, which he immediately sent back again. Having purchased fish, his intention was, to return to the vessel; but a sudden squall drove her out to sea; and, as the crew were all ill of the scurvy, Shelikoff concluded that they would make the best of their way to Ochotsk. He himself went to Bolshoiretsk, where he arrived on the 15th August, and bought three horses to travel by land to Ochotsk. While he was there, intelligence was received of the arrival of an English ship at the harbour of St. Peter and St. Paul, for which place he immediately set out. He arrived on the 23d of the same month, and was well received by the English; for his own account is as follows:

" When the English observed my arrival, some of them imme-
" diately came on shore in their boat. The captain and two offi-
" cers met us in a very friendly manner, and invited us to go
" on board their vessel, where they shewed me samples of their
" goods, and said that they had letters from the East India Com-
" pany to the commander of Kamtshatka, in which the company
" expressed a desire of opening a trade with Russia, and requested
" permission. I endeavoured to discover whence they came, and
" the course that they had steered; for they did not conceal their
" charts from me. I heard that they were from Bengal, which
" place they left, according to our stile, on the 20th March;
" sailed the 16th April from Malacca; arrived the 29th May at
" Canton; left it the 28th July, and arrived here the 9th August.
" They were three officers and a Portuguese. The crew consist-

" ed

" ed of Englishmen, Indians, Arabs, and Chinese: in all, 70.
" The vessel was built entirely of mahogany, had two masts and
" twenty-eight sails; was sheathed with copper to the gunnel,
" and mounted twelve guns.

 " After supper, Captain Peters with his officers conducted me
" on shore; but we deferred trading till the arrival of the com-
" mander of Kamtshatka, Baron Von Steinheil, who came on
" the 25th, and acted as interpreter, speaking the French lan-
" guage. They bound themselves to pay duty, according to the
" claims of government; I gave them a list of articles wanted
" here, for their future government; purchased goods to the
" amount of 6611 rubles; paid in part 1000 rubles, and gave
" bills upon Mosco at two months' sight, bearing interest till paid,
" at the rate of six *per cent.* On the 3d September I took leave
" of the Englishmen, who intended to sail the next day. On
" the 8th I arrived with my goods at Bolsheiretsk, where I im-
" mediately sold the whole for upwards of 10,000 rubles in
" ready money."

Captain Peters was afterwards wrecked upon Bering's island,
and only two of the crew saved (a Portuguese and a Lascar).
These travelled with me in the Autumn of 1788 from Ochotsk
to Yakutsk, in their way to St. Petersburg. The Portuguese told
me, that Captain Peters wanted to load his ship with copper,
which he had a notion that he might collect at Bering's or Copper island. In all probability he was misled by the exaggerated
accounts of the quantities of copper found upon those islands.

     CHAP.

CHAP. XX.

A Part of the Company fail, under Captains Hall and Saretſheff, for Ochotſk.—Intelligence received from Captain Billings and his Party.—Letter from Mr. Main to the Author, giving a brief Sketch of their Sufferings.—An alarming Earthquake.—La Flavia, a French Ship, arrives with ſpirituous Liquors and other Articles.

THE impoſſibility of entering the port of Ochotſk with our large veſſel compelled us to take the reſolution of laying her up in Kamtſhatka, and waiting (for our deliverance) the arrival of the tranſport-veſſel with the annual ſupply of proviſion for the peninſula. As, however, our company was too numerous to embark in one of theſe galliots, Captains Hall and Saretſheff determined to ſail with as many as they could take on board the Tſhernoi Orel. They were ready early in the month of July; but eaſterly winds prevented their departure until towards the latter end of the month, when they ſet ſail with an intention of exploring the Kuril iſlands and the coaſt of China to Ochotſk *.

Shortly after their departure, we received intelligence from Captain Billings, of his ſafe arrival at the river Angarka, after encountering the greateſt difficulties, and ſuffering innumerable

---

* They were prevented executing this undertaking by contrary winds, which detained them at the Kuril iſlands till late in Auguſt, when they thought it more adviſable to ſail direct for Ochotſk.

hardſhips

hardſhips from the Tſhutſki. I received a letter from Mr. Main, of which the following is a copy :

" DEAR SAUER,

" I ſhould think it a ſpecies of ingratitude to let ſlip an opportunity of writing to you, to inform you of our ſafe arrival at the river Angarka * on the 15th inſtant, after undergoing every thing that is bad during the ſpace of ſix months and two days ; ſuffering by the moſt violent froſts, without ſhelter from the bleak north winds ; owing to this barren country not producing the leaſt bit of wood, except when we fell in with rivers that afforded on their borders ſome creeping willows. We were therefore obliged to put up with the frozen meat of deer, and whales and ſea-horſe fleſh, raw ; and even with theſe the Tſhutſki fed us very ſcantily, not only almoſt ſtarving us, but at the ſame time robbing us daily before our faces. They alſo formed two plots, at different places, to murder the Captain and our whole party ; but God Almighty prevented their laying violent hands upon us ; and we have great reaſon to thank our Maker that we are now quite out of their power, and getting ready to ſet out for the Lower Kovima, for which place we depart to-morrow morning, accompanied by Mr. Bander †, whom we found here waiting the arrival of the Tſhutſki.

* The river Angarka is of no great extent. It commences near the ſource of the river Tſhaoon, or Tſhaun ; and, taking a contrary direction, flows into the Suchoi Annui, the latter diſcharging itſelf into the Kovima, oppoſite the village of Neizſhni.

† Mr. Bander is the Iſpravnik, or Captain of the diſtrict of Zaſhiverſk. His buſineſs on the Angarka was to collect tribute from the Tſhutſki. This gentleman's name has occurred frequently in the former part of this work ; but, having always mentioned it from memory, I have been led into a miſtake, in ſpelling it Bonnar, inſtead of Bander.

" I aſſure

" I affure you, that I very often curfed the hour wherein I left the Slava Roffie, having been obliged ever fince that time to bear with patience the abufes of the wildeft of favages, and expecting death daily.

" I have a great deal to communicate, but muft delay it till a future day, being too much confufed at prefent with the bufinefs of packing up, and joy at feeing our old acquaintance Mr. Bander, who travels at all times with a great ftock of good things; and, as we have had no fpirits now for thefe fix months, a little drop makes us very merry. Let me, therefore, conclude with affuring you, that I remain ever,

" DEAR SAUER,

" Your fincere friend,

" JOHN MAIN.

" The River Angarka,
21ft February 1792."

Several other letters mentioned that the Tfhutfki had deftroyed their meafuring lines, and their writing materials, and abfolutely prohibited the taking of any notes, or making remarks; which, however, without thefe violent reftrictions, was rendered impoffible by the fevere froft and driven fnow, which completely prevented their obferving the lakes from the land; and as they did not approach the fea-fhore any where, except the Bays of Melfhikma and Klutfhenie * (the latter was frozen at the time, and from hence their courfe was weft to the Angarka), they had not obtained any knowledge from their own obfervations of the fituation of Shalatfkoi Promontory, the Tfhaun Bay, or the direction of the coaft of the Icy Sea between the eaftern promontory and

---

* The mouth of this bay is at Captain Cook's Cape North.

the

the fartheſt place obſerved in 1787, in our excurſion to the Icy Sea; namely, 30 miles eaſt of Barannoi Kamen, the cape called by Shalauroff Peſoſhnoi Muys.

The letters mentioned, that Captain Billings's intentions were to go immediately to Yakutſk. He alſo deſired, in his papers to the Command, that I would make all poſſible haſte to join him at the above-mentioned place.

We were in daily expectation of the arrival of the tranſport veſſel; and our Company conſiſted of

Capt. Bering,
Mr. Bakoff,
Bakulin, } of the Expedition;
Robeck,
And myſelf,

beſides the commander of the harbour, Major Schmaleff, and his Aſſiſtant Enſign Roſtergueff. All the reſt of our neighbours were petty officers, ſailors, and Coſſacs.

As my buſineſs did not confine me to the harbour of St. Peter and St. Paul, I reſided chiefly at Paratounka, and made frequent excurſions on the chaſe with the Kamtſhadals, ſometimes for eight or ten days together, in the woods, and roving about the mountains at no great diſtance ſouth of the Bay of Avatſha. I ſaw bears in great numbers, wolves, foxes, and a few deer; but could only kill the former, as we had no dogs with us to run the other animals down. Hares alſo were in great plenty, but extremely ſhy.

On the 11th Auguſt, in the harbour of St. Peter and St. Paul, I obſerved a number of ſwallows flying about, apparently much frightened.

3

frightened. They were red breasted, a species never remembered to have been seen here; and the inhabitants immediately predicted some remarkable event; they were, however, only seen during the morning. The next morning, about five o'clock, we were alarmed by a violent shock of an earthquake, preceded by a rumbling noise, little short of thunder. The motion of the earth was undulatory for nearly the space of a minute. I was dressing myself, and was thrown down, which induced me to get out of the house as quickly as possible. The water in the bay was agitated like a boiling cauldron. The shock came from the northeast, and appeared to me to continue upwards of two minutes; but other gentlemen were of opinion that it did not last more than one. A sailor, one of the watchmen on board the ship, was thrown out of his hammock. At Paratounca it was more violent; the earth opened in many places, and water and sand were thrown up to a considerable height; all the buildings in the village were more or less damaged; one balagan was thrown down; some of the ovens (the only brick-work about the buildings) were also shaken in; and all the paintings, &c. in the church, except Captain Clerke's escutcheon, were thrown from their fastenings.

At Neizshni Kamtshatka the inhabitants were extremely terrified; nor could they explain whether the noise or the shock preceded. The situation of the town is on a neck of land formed by the discharge of the Raduga, a considerable river, into the Kamtshatka; the bed of the former was dry, and the inhabitants ran across it toward the mountains. They, as well as the cattle, were thrown down; and the continuance of the trembling was, according to their account, near an hour; the earth opened in many places, and sunk considerably in some. The volcano Klut-
shefskoi

fhefskoi emitted a vaſt column of black ſmoke; a noiſe like thunder ſeemed to iſſue from the bowels of the earth; the bells of
the two churches rang violently; and the howling of the dogs, and
ſcreams of the people, ſurpaſſed all deſcription, for the latter expected every moment to ſee the complete deſtruction of the town.
But when the ſhock was over, the loſt water of the river reſumed its former channel, and the inhabitants returned to their
dwellings. Not a ſingle brick chimney or oven was left ſtand-
ing. The altar of one of the churches * was ſeparated from it
about a foot, inclining a contrary way; and the greater part of
the balagans were thrown down.

It is remarkable, that the inhabitants of the village at the foot
of the burning mountain only heard the noiſe, and did not feel
the ſhock; nor did it croſs the mountains to the weſtern ſhores
of the peninſula.

Diſpatches from St. Peterſburgh to the commander of the port
announced the departure from France of a Ruſſian ſubject of the
name of Torckler on board his own veſſel, with a view of ſupply-
ing theſe diſtant parts of the world with proviſion and every
other neceſſary, and recommended the governor's aſſiſtance to
the ſaid Torckler. Towards the latter end of the month of September the veſſel arrived, a fine new ſhip of about ſix hundred
tons, copper-bottomed, and called La Flavia. Her crew conſiſted

---

* Churches and houſes throughout Ruſſia, in all ſmall towns and villages, are built of timber; the ſpars laid on one another; the ends notched to admit of their lying cloſe together, and the interſtices filled with moſs. The altars are detached ſpars at the eaſtern extremity of the church, built as cloſe to its body as poſſible. The top of the belfry at
Neizſhni inclines in one direction more than three feet over the foundation of the building.
It is about 40 feet high; and I think that the joiſts at the end of the ſpars are the only
means of preventing its fall.

of,

of, I believe, fixty men, befide officers. She carried the new French flag, and the officers wore the tri-coloured cockade. Mr. Torckler was the fupercargo only; the greater part of the cargo confifted of fpirituous liquors; and the captain and officers were in every refpect gentlemen and men of fcience. About the time when this fhip arrived, we were informed that the tranfport veffel from Ochotfk was driven on fhore near the river Itfha, between Bolfhoiretfk and Tigil. Captain Bering and Mr. Bakoff, therefore, went to that veffel to fee if they could afford her any affiftance; and I received the charge of the failors, &c. here.

CHAP. XXI.

The Peninfula of Kamtfhatka defcribed.

On the return of Captain Bering and Mr. Bakoff toward Chrift-mas, I made an excurfion to Neizfhni Kamtfhatka town, and re-turned to the harbour in the month of April; which trip, in ad-dition to others which I had before made, has enabled me to give the following account of the peninfula.

I fhall commence my defcription at the fouthern extremity, which the Ruffians call the Lopatka, latitude 51°, longitude 156° 40', eaft from Greenwich; a low point of land, widening and rifing gradually into mountains, barren and rocky, only pro-ducing here and there the creeping cedar and willow, to the ex-tent of 40 miles. Birch trees then appear in the inconfiderable vallies, which are replete with lakes and runs of water rufhing into the fea both eaft and weft. A clufter of mountains occupy the whole fpace from the Lopatka to latitude 53° 5', where, in the neighbourhood of the village Malka, they divide into two branches, one trending north north-weft; the other, which may be called the principal chain, leads north north-eaft. The place where the mountains feparate is the higheft land on the peninfula, and forms a barren ftony defert of 65 miles in length, in a di-rection north and fouth, and from 3 to 15 in width, pro-ducing in detached fpots brufh-wood, willows, and a very few fcattered and ftunted birch-trees. It is replete with fprings and

P p

brooks;

brooks; some of which uniting, and flowing south and south-west, form the Biſtrea; while others, at only a few fathoms diſtance, take an oppoſite courſe, and are the ſources of the river Kamtſhatka. At the end of this deſert, the mountains cloſe within a mile or two, and a foreſt of birch-trees follows to the village Apouſhinſk, where the river Kamtſhatka is navigable for ſmall boats to its diſcharge.

From this place the face of the country aſſumes the appearance of extreme fertility. The valley widens, and the ſpace between the mountains eaſt and weſt is at Virchni Kamtſhatka 40 miles. The ſoil is deep and rich, compoſed of black earth, mixed with fine black aſhes from the burning mountains, and fine iron ſand, which adheres to a magnet, and forges well with bar-iron, but uſed alone is very brittle.

The productions of nature are, a ſmall kind of wild black cherry (tſheromka), in great abundance; the wood of which, being particularly hard, is uſed by the Kamtſhadals for their guiding ſticks to the ſledges: the thickeſt trees that I have ſeen are nine or ten inches in circumference. Firs, common pine and larch trees of extraordinary ſize, with birch, poplar, aſp, and mountain-aſh, clothe the mountains to their ſummit. The underwoods are, currant, dog-roſe, hawthorn, alder, and buſhes producing berries.

The climate is very different from that of the ſouthern and northern parts of the peninſula, the valley being completely ſheltered from the ſea-breezes that chill the air in other parts, and prove a great check to vegetation, which commences here in the month of March. The ſcenery is beautiful beyond deſcription,

the

the river meandering through the midſt of the valley, from 50
to 250 yards wide, and from eight to 15 feet deep, and being
replete with trout and every ſpecies of ſalmon in the ſeaſon.   This
valley is 180 miles in length, frequently opening proſpects of the
Tolbatſhinſk, a lofty double-headed mountain, conſtantly emitting
an immenſe column of black ſmoke; while the ſecond volcano,
Klutſhefskoi, towering to an incredible height, illuminates the
clouds with its blaze, and affords a view awfully grand.

Twelve verſts below Virchnoi Kamtſhatſkoi Oſtrog, is the village called Milkovoi, inhabited by farmers ſent from Siberia at
government expence, and poſſeſſing particular emoluments, for
the purpoſe of growing corn and pulſe to ſupply the country.
They live uncontrolled, however, and find it eaſier to accumulate
wealth, by acting as retailers for the merchants of Kamtſhatka,
and going themſelves on the chaſe for ſables, &c. than in pur-
ſuing the more toilſome labour of cultivating the earth, which
they neglect.   Yet they keep gardens that produce very fine
cabbages, potatoes, turnips, carrots, cucumbers, &c.; they alſo
grow buck-wheat and rye for their own uſe, which yield abundantly; and I am inclined to think that, had they a proper in-
ſpector to ſuperintend their buſineſs, they might with eaſe grow
corn enough of every kind to ſupply not only the peninſula, but
all the neighbouring country, Ochotſk, &c.   Hemp grows remarkably well, which, however, I think there is no need of cultivating; for the nettle ſeems equally to anſwer every purpoſe.
The Kamtſhadals and Ruſſians make ſewing thread of it, and fa-
bricate from it their fiſhing nets, which ſerve them, if uſed with
caution, and properly dried, four or five years.   The proceſs of
preparing it is nearly the ſame as that for hemp, but I think leſs
troubleſome; the nettle grows to the height of ſix and ſeven
feet;

feet; the fibres are much finer; and thread of equal thickneſs is ſtronger than that made of the imported hemp.

At Tolbalſhinſk the mountains are broken and barren; they encroach upon the valley, and conſiderably leſſen its width. Storms are frequent between this volcano and that of Klut-ſhefskoi, but never reach the neighbourhood of Milhovoi, and the trees are conſiderably leſs in ſize; but the country continues fertile as far as 30 verſts north of the village Klutſhefskoi, which is alſo a colony of Siberian peaſants for the ſame purpoſe as thoſe at Milkovoi, and who act in the ſame manner. Their proceedings are in ſome reſpects juſtifiable; for the magiſtracy at Neizſhni exact the ſame payment from them as from the reſidentiary merchants. The court of juſtice conſiſts of a burgo-maſter, four members, or rathmen, a ſecretary, writers, and watchmen; receiving a ſalary for the time they are in ſervice: and frequently an expenſive deputation is ſent to Tigil, Bolſhoiretſk, Virchni, and ſuch places as are reſorted to by theſe pedlars; ſo that it is a matter of doubt with me, whether the culture of the earth would render any harveſt ſufficient to anſwer the payment of their claims. In ſome years they amount to 18 or 20 rubles, and in others half as much more.

As you approach the north, the ſeverity of the climate increaſes; the ſoil becomes ſandy and ſtoney; and the vegetable productions are ſtunted and weak. The iſthmus is ſituated in latitude 59° 20', and the diſtance from ſea to ſea is here about 40 miles. The wideſt part is from Kronotſkoi Noſs to the river Itſha, about 220 miles.

I have

I have already mentioned the fituation of the town Neizfhnoi Kamtfhatka, which contains 80 ifba's, or houfes, with two churches; and its number of inhabitants, including children, 548 fouls; latitude 56° 33′.

The weftern coaft of Kamtfhatka is uniformly low and fandy, to the diftance of about 25 to 30 miles inland, where the mountains commence. It produces only willow, alder, and mountain-afh, with fome fcattered patches of ftunted birch-trees. The runs of water into the fea from the mountains, do not deferve the name of rivers (except the Bolfhoia Reka), though they are all well ftocked with fifh from the fea in feafon, as trout and different fpecies of falmon. They are generally at the diftance of 15 to 20 miles from each other. The Itfha and Tigil are the moft confiderable; and neither of thefe have a courfe, with all the windings, of more than 100 miles.

The fea is fhallow to a confiderable diftance; and the commanders of the tranfport veffels, who never lofe fight of the expofed coaft if they can help it, judge of their diftance from land, in foggy weather, by the foundings, allowing a fathom for a mile; nor is there at the entrance into any of the rivers more than fix feet at low water, with a confiderable furf breaking on the fandy beach.

The villages on this coaft are, Tigilfk, Itfhinfk, and Bolfhoi-retfk (fituated on the Tigil, Itfha, and Bolfhoia rivers). Of thefe the former is the moft confiderable, containing 45 wooden houfes, and one church. The Ruffian charts place it in latitude 57° 55′. This, which they call a fortified town, is furrounded by wooden palifades, and was built in 1752. The number

ber of inhabitants are 338, including women and children. It-shinsk also contains a church, and about 10 houses, with 50 inhabitants. Bolshoiretsk contains 37 houses, and the total number of inhabitants are 235. Beside these, there are eight inconsiderable villages, containing each three or four houses, on the west coast.

The eastern coast is composed of mountains, rocks, rugged cliffs, and bold promontories, replete with inlets, and the appearance of such. Their entrance, however, is blocked up by reefs of rocks, the openings of which are only to be entered by the boats of the natives. Immense masses of stone are scattered out at sea to the distance of one, two, and three miles; some of them being only discernible by the breakers, while others tower to a considerable height. The depth of the sea varies much, and suddenly, from 30 to 90 fathoms, and more. Earthquakes are frequent, and sometimes very violent.

The only harbour for ships on the whole peninsula is the Bay of Avatsha *, which is probably the safest and most extensive in the world. I shall exert my utmost ability to describe this place; but fear that I shall scarcely be able to do it justice.

I will suppose myself approaching the coast from the south-east. When first seen, it appears strait and uniform, without bays or inlets; the land rising into moderate mountains, backed by such as are more lofty. Three of them, apparently united,

---

* " The term Bay, properly speaking, is rather inapplicable to a place so completely sheltered as Avatsha; but when it is considered how loose and vague some navigators have been in their denominations of certain situations of sea and land, as harbours, bays, roads, sounds, &c. we are not sufficiently warranted to exchange a popular name, for one that may perhaps seem more consistent with propriety." *Cook's last Voyage.*

I

are very confpicuous to the north of the Bay; the fartheft, or the moft weftern, is the higheft, and is conical; the next is a volcano, diftinguifhable by a column of fmoke iffuing from its fummit, which is broken; the third prefents feveral flat tops, lowering, and trending eaft, from which a narrow and lofty broken and irregular point of land extends about 15 leagues, terminating in a promontory called Sheeponfkoi Nofs. South of the bay are two remarkably lofty mountains; that neareft the entrance (Vi-luitfhefkoi Sopka) is formed like a fugar loaf; the other, Apal-fkoi, is far inland, not fo lofty, and is flat on the top. On getting well in with the land, it is high, craggy, and broken, prefenting the appearance of inlets. When about the latitude of 52° 45', and longitude 159° 15', the entrance into the Bay of Avatfha difcovers itfelf, bearing north-weft by weft; fouth of which, at the diftance of about four miles, is a fmall round ifland, compofed of high pointed rocks, called Staritfhkovoi Oftrov. The north cape is a bluff head, with a light-houfe on the top of it, refembling a centinel's box. From this cape eaftward, to the diftance of three miles, breakers are difcernible over hidden rocks, which extend to the fouth about half a mile. Within the channel, are three detached needle rocks near the north fide : on the oppofite fhore a fingle one remarkably bulky, the top of which is nearly flat. Soundings leffen from 40 to 12 fathom, over a ftoney bottom; and 10 fathom in the channel, fand and mud.

The entrance is in latitude 52° 51', longitude 158° 48, whence Shëeponfkoi Nofs bears eaft north-eaft, diftant about 17 leagues. It forms a channel in a direction north north-weft four miles deep; the breadth is three miles in the wideft, and two in the narroweft part; both fhores rocky; the fummits covered with birch trees, mountain-afh, and hawthorn. Having paffed this
                                                                channel,

channel, you enter a moſt magnificent baſin about 25 miles in circumference, completely land-locked, and every where (except the north-weſt extremity) high, and covered with trees. As we advance in the baſin, commodious harbours open : to the eaſt, Rakivinoi, about three miles deep, and three quarters of a mile wide. The ſouth cape is a lofty perpendicular maſs of ſtone. Shoal water over rocks extend from the ſhore into the bay, about 50 fathoms from ſouth to north, which makes the entrance diffi-cult. The north cape is a high rocky ſhore, with ſome rocks that are detached ; but theſe are viſible, and not of any extent ; the depth within is from 13 to 3 fathoms.

The little harbour of St. Peter and St. Paul is to the north : its ſouth bluff cape is extremely conſpicuous ; and the buildings upon the ſpit of land before it are clearly diſcernible. This har-bour is in every reſpect convenient for giving ſhips all kinds of repair, as alſo for obtaining fire-wood and freſh water. If all its good qualities are conſidered, I think it may be pronounced the moſt convenient in the world. Six or eight ſhips of war might be conveniently moored in it head and ſtern : the only inconve-nience (if it may be ſo called) lies in the amazing toughneſs of the ground ; for if the anchor be heavy, and out any time, it will probably be found neceſſary to heave a ſtrain on the ſhip to weigh it. The ſouth of the harbour is bounded by a narrow neck of low land running out from the eaſtern ſhore in a weſtern direction, covered with wooden houſes and balagans ; at the ex-tremity of which is the entrance, 38 fathoms wide, and ſix and a half deep : ſhips may paſs ſo cloſe to this point of land, that a man may leap on ſhore. The weſt ſide is a projecting narrow mountain of moderate height trending to the ſouth, terminating in a bluff head, upon which is a battery of three guns, and a flag.

From

From this extremity a fhoal extends fouth about 100 fathoms. The north, which is the head of the harbour, is bounded by a valley, in which are the government magazines, barracks, and the dwelling houfes of the commander and chief inhabitants. To the eaft, it is bounded by lofty mountains covered with wood to the very fummit ; namely, birch, mountain-afh, hawthorn, dwarf-cedar, rofe-bufhes, &c. On this fide are feveral fprings of very pellucid water rufhing down the mountains into the harbour, and thefe are very convenient watering places.

The bay is bounded to the north-weft by the extenfive plains of Avatfha, where two rivers difcharge themfelves into it ; namely, the Avatfha and Paratounca. At the eftuary of the former, the Kamtfhadals, who formerly refided in the harbour, have their prefent habitations ; while the troops of the garrifon poffefs their late dwellings.

The harbour of Tareinfk opens to the weft : the entrance is about one mile in length, when it fuddenly turns to the fouth-eaft, extending twelve miles in length, and three in width : the depth is from fix to eight fathoms, mud and fand. A narrow neck of high land at the bottom, like an artificial partition, fe-parates it from the fea. This harbour, although extremely con-venient in its conftruction, is not fo in its fituation ; as an eafterly wind is abfolutely neceffary to bring outward bound veffels into the bay of Avatfha where it is quite contrary ; befides, it is ex-pofed to the north and north-weft winds, which blow over the plains of Avatfha right into it, and keep it blocked up by ice till late in the feafon.

Near

Near the mouth of this harbour, on its north-weſt borders, is a valley forming a plain of about one mile and a half ſquare, well wooded with good ſized birch trees : a ſituation which Major Behm thought the moſt eligible of any hereabouts for a town. North and ſouth are lofty mountains of eaſy aſcent, covered with trees to their ſummit. The valley is bounded to the weſt by a freſh water lake of about 15 miles in circumference, well ſtocked with fiſh all the year through ; while its borders abound in different kinds of berries, ſarana, tſheromtſha (a kind of wild garlick), and a variety of pot herbs. This lake is the chief ſource of the river generally known by the name of the Paratounca, of which I ſhall ſhortly give an account, as alſo of the other rivers. The remains of numerous villages in the vicinity of this lake ſtrongly indicate the former populouſneſs of theſe parts ; but they are at preſent overrun with bears, wolves, and hares.

The fiſh of the bay of Avatſha are, cod all the year through ; thornbacks, flounders, and halibuts, as ſoon as the ice begins to break ; whitings are caught all the winter by the boys and girls, who make a hole in the ice, lie flat upon it, and look into the water, holding in it a horſe-hair nooſe on the end of a ſtick, which they get round the fiſh, and by theſe means haul them up very faſt. Herrings and ſmelts are the firſt paſſage fiſh that appear (the former in immenſe ſhoals towards the latter end of April), and they remain till the beginning of June : their numbers, indeed, are incredible. In Cook's laſt voyage Captain King ſays, " The people of the Diſcovery ſurrounded ſuch an amazing " quantity (of herrings) in their ſeine at one time, that they were " obliged to throw out a very conſiderable number, leſt the net " ſhould be broken ; and the cargo they landed was ſtill ſo abun-

" dant,

" dant, that, befides having a fufficient ftock for immediate ufe,
" they filled as many cafks as they could conveniently fpare for
" falting ; and, after fending on board the Refolution a tolerable
" quantity for the fame purpofe, they left behind feveral bufhels
" upon the beach."

On the 7th June, in the inner harbour of St. Peter and St.
Paul, I obferved, at the flood tide, a confiderable number of herrings fwimming round in circles of about a fathom in diameter.
Seeing them continue in this particular manner, I approached
very near them, and remarked, in each of the circles, one fifh
very clofe to the ground, upon the weeds, and apparently without motion. I could not account for this peculiarity in their
fwimming, but thought that the weeds about the herring in the
middle became of a very lively yellow colour. When the tide
ebbed, and left thefe places dry, all the weeds, ftones, fticks, &c.
were covered with fpawn about half an inch thick, which
the dogs, gulls, crows, and magpies, were devouring with great
avidity. Thefe fhoals of herrings, which are purfued by feals,
cod, &c. come in fpring and in the autumn ; there is, however, a
confiderable difference in their fize ; and I believe the fpring glut
are the largeft fifh. The natives and other inhabitants enfnare
a great quantity in autumn for their dogs.

The oil extracted from herrings is very pellucid and fweet ; it
preferves birds' eggs all the year quite frefh and good, as does
alfo the oil of feals, whales, &c.

Herrings no fooner difappear, than immenfe fhoals of falmon
pufh into the bay and up the rivers; the firft is the fmaller fort,
called the gorbufhka (or hunch-back), from a remarkable protu-

tuberance

berance which diftinguifhes this fpecies. They are in great perfection about four or five weeks; and are no fooner in a ftate of
decay, than another and larger fpecies follows. Thefe are fucceeded by other forts, all of the falmon kind, until the end of
September. I could not diftinguifh fome forts till they were
boiled, when the palenefs or rednefs of the flefh indicated a difference. The names of the different fpecies are, fiomga, tfhavitfha
(both very large), gorbufhka, kaiko, krafnaia (red), and belaia
(white). They pufh up the rivers, and get into the lakes, where
the two latter continue all the winter, but get extremely deformed,
crooked-backed, thin, and covered with red blotches; the upper jaw
extends beyond the under jaw, and bends over it; the mouth is
full of very large teeth, and the flefh is flabby. Here are alfo
great varieties of crabs, fea eggs, prawns, mufcles, cockles, and
the fmall pearl oyfter.

The plains of Avatfha, which bound the bay to the north-weft,
extend 18 miles from eaft to weft, and 35 miles from north to
fouth, producing at their northern extremity only a few patches
of birch, poplar, and alders; all the reft are marfhy grounds,
overgrown with rufhes and coarfe high grafs, with fome fpots of
oziers and alders; the refort of myriads of water fowl, fwans,
geefe, an amazing variety of the duck kind, and I believe every
fpecies of fnipe. The river Avatfha runs in feveral branches
through this plain. Towards its northern extremity are an immenfe number of cold fprings, that form feveral bafins of water,
with fmall runs, uniting in a rivulet, which has a courfe towards
the fouth of two miles, and empties itfelf into the river generally
known by the name of the Paratounca, but which is called by
the natives Ilmitfh; and the original Paratounca is the run from
the fprings above mentioned, oppofite to the difcharge of which
the village of that name is fituated. Thefe fprings do not

freeze

freeze in winter, during which feafon they are frequented by fwans, geefe, and feveral fpecies of duck, particularly the fafka, or duck with a melodious note, which has induced me to ftile it mufical. Here they find an abundance of food, and the pulpy root of an aquatic plant which, in its appearance, exactly re-fembless the olive, but is like the chefnut in flavour: I ufed it as a vegetable, and thought it better than any production of the gardens: the natives call it the farana of the geefe.

With regard to rivers, the Kamtfhatka is the only one of any confequence. Its fource I have already defcribed: it flows nearly north to Neizfhni Kamtfhatka, where it turns to the eaft fouth-eaft about 25 miles, and empties itfelf into a large but fhallow bay formed by the Kronotfkoi and Kamtfhatfkoi promontories; its difcharge is extremely fhallow, not exceeding eight feet at high water, and the breakers are very violent with an eafterly breeze. This, however, is the only navigable river on the peninfula.

The Bolfhoia Reka has only a courfe of 20 miles. It is formed by the union of the Byftrea with the Natfheke, a little below Bol-fhoiretfk: the former has its rife from the fprings near the fource of the Kamtfhatka, and takes a fweep from fouth to weft; the latter commences a little fouth of the village Natfheke, and flows nearly weft about 100 verfts: neither are navigable, though, during the fpring flood, the natives fometimes venture down them in their canoes, but with great difficulty, owing to ra-pids, &c.

The Avatfha has alfo an interrupted and unnavigable courfe of 70 verfts in a direction eaft fouth-eaft. The inhabitants of the
village

village Koriatſk, 20 miles up this river from its diſcharge, paſs up and down it in their canoes, hauling them over the flats.

The Ilmitſh, commonly called the Paratounca, has its riſe from near the Viluitſhiſkoi Sopka (called in Cook's laſt voyage the Paratounca); and from a lake already deſcribed near the Tarein-ſkoi harbour, it makes a circuit of 85 verſts, and diſcharges itſelf into the bay of Avatſha, only three miles in a direct line from its ſource: it is navigable for boats all the way; but I was fourteen hours in traverſing the whole of this river, from the lake, in a canoe. The villagers of Paratounca go on the chaſe of deer, ar-gali, bears, &c. about the ſource of this river, by paſsing down the ſtream into Tareinſki harbour, and hauling their canoes over the plain already mentioned into the lake. An immenſe number of rivulets from the mountains flow to the eaſt into the ocean, but none of them are either remarkable, or have their banks in-habited.

Here are no lakes of any extent: the names of the principal ones are, Oſernoi, about 40 miles from Cape Lopatka; Kronot-ſkoi, 20 miles ſouth-eaſt of Tolbatſhinſki volcano; and another of leſs extent, ſituated about 40 miles north of Neizſhni Kamt-ſhatka, called Nerpitſhi: the natives ſay that they are replete with fiſh; and *tradition* relates, that the fiſh of theſe lakes had two heads, or that they poſſeſſed legs; and, being ſacred to ſome deity or demon, thoſe who preſumed to enſnare them were pu-niſhed with misfortunes: ſome of the natives, however, ſeem to doubt the *truth* of this, while others ſtill firmly believe it.

Hot

Plate XII.

A View of the Ozernoi Hot Springs of Kamtchatka.

Hot fprings are very numerous, and feem fcattered all over Kamtfhatka; but thofe in the following places are the moft remarkable:

Opalfki, or Ofernoi, fituated nearly midway between the Lopatka and Bolfhoiretfk, about 15 miles fouth of the Kamtfhadal village of Yavinfk, furrounded by mountains, and at no great diftance from the volcano of Opalfk. They occupy a valley of confiderable extent, and are fcattered to the diftance of fix miles, fome parts of which produce detached birch trees, the fweet plant, &c.; but in general the foil is barren, compofed of different coloured marl, and large ftones which appear to have been fcattered by eruptions of fome volcano. The largeft hot fpring is at the foot of one of the mountains; and we heard the noife that it made at the diftance of near a mile before we came to it. It is about fix fathom in circumference, boiling up to a confiderable height; the middle appears like a cauldron; and a piece of beef placed in it was very well boiled in a fhort time: all around, it bubbles up between large ftones; it then divides into two ftreams, which defcend over ftones, and unite at the bottom with a fmall rivulet formed by the other fprings to the north: they flow a little way to the fouth, then turn weftward into the lake Ofernoi. About the border of thefe fprings, and the rivulet which they form, we obferved petrified, or rather calcarifed, foliage of the fweet plant, birch leaves, fticks, &c. of a beautiful whitenefs; but fo extremely delicate in their texture, that we could not preferve any, even in cotton; for they mouldered to duft. The Kamtfhadals fuppofe this to be the habitation of fome demon, and make a trifling offering to appeafe his wrath; without which, they fay, he fends very dangerous ftorms. Our naturalift and Mr. Varonin, who afcended to thefe fprings in 1790, experienced

14   a whirl-

a whirlwind, which tore their tent, and fcattered its contents about, many of which were never found again. Afhes were fcattered upon the fnow about four inches deep, refembling coarfe gunpowder, probably from the volcano Alaid (a folitary mountain in the fea, fituated about 20 miles fouth-weft of the Lopatka), which burns violently at this time (February 1793). It has at various intervals emitted fmoke ever fince 1790. The oldeft inhabitant does not remember its having done fo before, although tradition informs them of its violent eruptions.

Toward the fource of the Byftria, near the village Malka, are hot fprings, a little way up the afcent of one of the mountains, which boil out of the earth in two or three places about a foot wide. Similar fprings are feen near the village Natfheke, but more extenfive, and forming in their run feveral convenient bathing-places. Thefe have a fulphureous fmell; and the ftones taken from the bottom of the openings, where the fprings appear, are covered with a fhining thin coat, which refembles filver at firft, but gets dull and of a dark colour after it has been fometime expofed to the air: the furrounding earth, to the diftance of 20 fathoms in every direction, is warm, replete with empty fhells like thofe of fnails, and a tranfparent glutinous fubftance; as alfo with fpots of loam, whereon any thing heavy being thrown finks immediately. South of thefe fprings, about the diftance of 30 verfts, at the fource of the river Natfheke, is a fandy level fpot, with feveral hot fprings, the water of which is faid to be brackifh.

At the diftance of 12 verfts from the village of Paratounca, in a direction north-weft, is the difcharge of a deep rivulet of warm water, called Klutfhevoia, navigable for canoes three verfts upwards.

wards. It fprings from feveral hot water lakes in an extenfive plain; one of which lakes is about 100 fathoms long, and 7 fathoms wide; very convenient for bathing near the fhore, but the middle very deep, and extremely hot. About 20 fathoms from this is another, about 5 fathoms by 7, but exceffively hot: a body of boiling water iffues through a fquare hole in a ftone at the eaftern extremity; and it has a run into a cold water fpring, fo narrow, that you may ftand with one foot in each. Ulcers, old and frefh wounds, are reputed to heal from bathing in this water. I ufed it for tea, but the flavour was not very agreeable, being fomething like that of alum. The hot fprings of Shumat-fhik are fituated 90 verfts north of the bay of Avatfha, and flow into Kronotfkoi Bay. There are feveral others, but of no note.

The following are the principal volcanoes:

Opalfk: I have defcribed this mountain as feen from the fea. Its fituation is near the hot fprings; but its emiffion of columns of fmoke is of very recent date, and they are not conftant; nor has it ever been obferved to blaze.—Viluitfh, or Viluitfhifkoi Sopka: this feems now completely extinguifhed:—Avatfha, 25 miles north of the bay, conftantly fends forth a body of fmoke from its fummit; as does alfo Tylbatfh, and Klutfheffkoi, or Kamtfhatfkoi Sopka, both fituated near the river. Tylbatfh (frequently written Tolbalfhinfk) is one of the mountains that conftitute the eaftern chain; but projects confiderably towards the river. It is more lofty than the reft, and has a pointed top. A little way down it, a fharp ridge ftretches away to the north; from this ridge, and the fide of the mountain where it joins, the fmoke iffues. I have obferved, in a clear night, a reflection over it, refembling the Aurora Borealis. Klutfheffkoi volcano may be reckoned among

R r

the

the higheſt peaks, I believe, in the world. It is ſituated 175 miles weſt of Bering's iſland, from which, however, it is diſtinctly ſeen in clear weather at the time of the ſun's ſetting: at leaſt, I am aſſured of this by ſeveral Kamtſhadals who have been on the iſland. This volcano is frequently ſubject to eruptions: in 1789, on the 20th November, a great noiſe preceded an earthquake; flames burſt forth, with diſcharges of ſmall ſtones and aſhes: the trembling of the earth and the noiſe continued, more or leſs, till the 23d, when it abated conſiderably; but on the 15th February '1790, it again reſumed its former violence until the 21ſt: all this time earthquakes were felt two or three times in the courſe of every 24 hours.—Shevelutſh is 80 verſts north of Klutſheffſkoi: this burnt formerly; but now it ſeldom happens that ſmoke iſſues from it: this volcano is the ſource of two rivulets, the Ilt-ſhutſh and Bakus, both of which flow into the Kamtſhatka.

The number of inhabitants may be ſtated as follows, men, women, and children:

At the town (Neizſhnoi)	548
Oſtrog Tigil	338
Virchnoi	226
Bolſhoiretſk	235
And at the harbour of St. Peter and St. Paul	85
Coloniſts	255
Ruſſians	1687
Kamtſhadals	1053
Total	2740

Of the natives 351 males only pay tribute, or, rather, are living on the liſt of thoſe who are tributary, according to the reviſion

made

made during the government of Mr. Reinikin, who fucceeded Major Behm. The revifor (in 1784), by fome unaccountable miftake, has frequently noted the name of one and the fame perfon as the inhabitant of two or three villages; and from each village the tribute is exacted for this man. They have made repeated remonftrances, but in vain: this impofition, however, is the leaft of their fufferings. They are compelled to pafs the greater part of their time in procuring neceffaries for their vifitors. The governor makes his annual circuit round the peninfula, and receives a prefent from every individual; the captain of the diftrict goes his rounds twice; different deputations from the courts of juftice, foldiers on furlough, couriers, &c. all travel at the expence of the poor native, who is compelled to keep an extraordinary number of dogs for their conveyance. Government horfes are quartered at each village, and the inhabitants muft provide a ftock of hay for them. Thus the Kamtfhadal fcarcely finds time to collect a fupply of food in the fifhing feafon for his own family.

In 1768 the fmall-pox carried off 5368 of the inhabitants; and fince the departure of Major Behm, the court of the interior (Zemfkoi Sud) has difcovered, that the Kamtfhadals are indebted to government the whole tribute for the unfortunate fufferers by that diforder, and lay claims at prefent for the debt. The natives produce receipts; but are told, that an ukafe from Irkutfk claims the payment. They appointed a delegate to lay their grievances at the feet of their fovereign; he, however, only reached Irkutfk, when he was promifed redrefs, and fent back again: he returned laft year, and is the chief of Shapinfki village, a very intelligent man, and, I thought, very likely to help me to

fome

some information as to their former customs and religion, which are now quite abolished; nor is their language pure.

He told me, that the Kamtshadals called themselves Itolmatsh (he says they are the Aborigines of the place), and the descendants of Newsteach or Newchtshatsh, and that their God was New-steachtshitsh. Koutka is his intelligent spirit, the messenger of vengeance to their tormenting demons, and of rewards to the spirits of benevolence: he travels about in an invisible carriage drawn by flying animals resembling mice, but smaller than the human mind can conceive, and swift as a flash of lightning. " Our Sorcerers (said he) were observers of omens, and warned us of approaching dangers, to avert which sacrifices were made to the demons: we were then wealthy, contented, and free." He continued his discourse thus as nearly as I could translate: " I think our former religion was a sort of dream, of which we " now see the reality. The Empress is God on earth, and her " officers are our tormentors: we sacrifice all that we have to " appease their wrath, or wants, but in vain. They have spread " disorders among us, which have destroyed our fathers and mo- " thers; and robbed us of our wealth and our happiness. They " have left us no hopes of redress; for all the wealth that we " could collect for years would not be sufficient to secure one " advocate in our interest, who dares represent our distress to our " sovereign."

They are an honest and hospitable race of men, extremely fond of music and of brandy. One of them, who constantly accompanied me in my aquatic excursions, and expended every farthing of his money in brandy, I one day saw coming to my habitation; and, to tempt him, I hid myself in an adjoining

room,

room, leaving a glaſs of brandy upon the table, and a bottle half full cloſe to it, with ſome ſea-biſcuit. He came in, ſaw nobody, and called me, but obtained no anſwer. Upon which he advanced to the table, and ſmelt to the glaſs: "It is brandy," ſaid he, "but I will not drink; and the bottle half full; well, I won't "taſte you; but I'll go and ſeek maſter, and ſcold him for "leaving you in this manner. I'll juſt ſmell again, and go."— I ſtepped out of window into the garden, and went to meet him; when he accoſted me in the following manner: "I have been "into your room and ſaw a glaſs full of brandy; perhaps you "won't believe me, but indeed I did not taſte it."—"I dare ſay "you did."—"No, by G—, I did not: I knew you would "not believe me; but a Kamtſhadal will never take any thing "without permiſſion."—"Well, I muſt believe you; will you "come and drink it?"—"Yes, that I will; but I wanted to ſcold "you for leaving it ſo."

They have long ſince adopted the manners of the Ruſſians, and profeſs the Greek religion. Of their former cuſtoms there only exiſt their laſcivious dances, and their impure language, with part of the dreſs. They entertain the greateſt veneration for the memory of Major Behm, under whoſe command they enjoyed the protection of a father. At that time Kamtſhatka was governed merely by the major and his aſſiſtant, Captain Shmaleff, without any other court of juſtice. The revenue of this peninſula was then 40,000 rubles annually, ariſing from the tribute of the natives, the profits on brandy, and the duty on furs; which ſum was remitted to the chancery of Ochotſk: a trading expedition in two or more open boats was yearly undertaken by the natives, accompanied by the prieſt of the peninſula, to the Kuril iſlands; from whence they obtained ſea-otter ſkins of a ſuperior quality,

quality, feveral Japanefe articles, and wrought filks. The merchants who vifited the peninfula brought other neceffaries, and
hard money for the articles of the natives. Brandy not being
confidered as an article of trade, but a government concern, the
fale was prohibited; and, the commander or his affiftant infpecting the tranfactions of the merchants, prevented fraudulent proceedings on both fides. About the latter end of the year 1779,
or the beginning of 1780, Major Behm returned to St. Peterfburg.
Major Reinikin fucceeded him in the command, and wifhed to
introduce among the natives the culture of the earth, but could
not fucceed; he brought potatoe feeds with him; firft grew
them in his own garden; and from thence they were abundantly
diftributed all over the peninfula, with every fpecies of garden
vegetables, which are cultivated with great fuccefs by the Ruffian
inhabitants.

In the year 1783 a mandate from her Imperial Majefty *pro-
claimed* Neizfhni Kamtfhatka a city; ordained it the feat of government of the country under the chancery of Ochotfk; offered
privileges to fuch merchants as chofe to become burghers; and
inftituted courts of juftice, eftablifhments better calculated to govern 300,000 men than 1500, which is about the number of
male inhabitants. The governor was denominated Gorodnit-
fhik (mayor); and his eftablifhment confifts of a fecretary and
writers; a Kaznatfheftva (exchequer) for the receipt of the revenue, and payment of officers; a Zemfkoi Sud (court of the
interior), of which the Ifpravnik, or captain of the diftrict, is prefident, and in this court one of the natives is a member to reprefent the whole body; with a magiftracy to regulate mercantile
concerns; as already mentioned.

The

The falaries allowed by government to the different officers are as follow :

The Gorodnitfhik - -	600 rubles.
His Secretary - - -	300
The Kaznatfhae - -	400
Ifpravnik - -	400

other officers lefs in proportion ; fome of the writers having only 24 rubles per annum.

Price of Articles 1793.	Kazan.	Kamtfhatka.
Linen for fhirts per arfheen *	18 cop. *	120 cop.
Boots per pair -	3 rub.	12 to 18 rub.
Thread ftockings do.	125 cop.	4 to 5 rub.
Soap per lb. -	6 cop.	60 to 100 cop.
Candles do. - -	8 cop.	80 to 100 cop.
Tea do. - -	2 rub.	12 rub.
Sugar - -	50 cop.	3 rub.
Leaf tobacco per lib. -	5 cop.	3 rub.
Rye flour, per pood * -	50	500 cop.
Wheat do. - -	60	800 cop.
Rice per lib. -	10	100 cop.

From this ftatement of prices, the impoffibility of an officer living upon his falary will plainly appear ; he is therefore compelled to find out fome method of increafing his income, at the expence of the poor natives.

One of the captains of the diftrict, who came here with his wife and family, finding himfelf extremely diftreffed, appropriated

* A pood is 40 lb. Rufs, or 36 lb. Englifh. 100 copeaks make a ruble ; a copeak may be reckoned little more than a farthing fterling. 9 arfheens make 7 yards.

the tribute of one year to his own ufe, and wrote a letter to the Emprefs; ftating, that the feverity of the climate, the prices of every article of life, and the wants of his family, had compelled him to make ufe of the tribute, confifting of fuch a number of fables and fox fkins for their backs and bellies, which he rather chofe to do than rob the poor natives (the only alternative). He requefted her pardon, and an appointment where he could live upon his falary; and the induftry of his family (of no benefit in Kamtfhatka) might help to repay the amount of the articles that he had appropriated to his own ufe. The Emprefs ordered the governor to give him fuch an appointment, and pardoned him on account of the good reafons that he affigned; but this pardon was not to be regarded as a precedent; for fuch mercy was not to be extended to any future perfon who fhould dare to act in the fame manner.

The magiftracy receives its income from the burghers and me-fhanin. The latter are privileged pedlers (and the colonifts are of the number); the former are divided into three claffes, ac-cording to the extent of the capital that they give in, upon which they pay one per cent.

French brandy is now regarded as an article of trade; and a fpurious fort is carried about the villages of the natives, who are very fond of it, and pay for it at the rate of one ruble per glafs.

The Kamtfhadals and refidentiary Ruffians employ themfelves during the fummer in catching fifh; drying fome, and falting others for a winter fupply for themfelves and their dogs: in the autumn, in making hay for their cattle, collecting berries, the

6

fweet

fweet plant, and kiprey; the former is purchafed by government for the diftillery of brandy, at three and four roubles the
pood when prepared and dried. In the fpring they collect
birds' eggs about the marfhes, and particularly among the rocks
at the mouth of the bay of Avatfha: thefe they preferve all the
year with oil, as already mentioned.

CHAP. XXII.

The La Flavia departs for Canton.—A Galliot arrives, and conveys the Party to Ochotſk, after ſome danger from a Leak.—The Author, with the firſt Party, ſets forward for Yakutſk; at which, however, he arrives alone, after encountering much Diſtreſs, and leaving his Companions and his Baggage behind on the Road.—Actively aſſiſted by the Commandant and Captain of the Diſtrict.—The Amoor River deſcribed.—Arrive at Irkutſk.—Sketch of Captain Billings's Expedition acroſs the Land of the Tſhutſki.—Arrival at St. Peterſburg.

ON the 1ſt of June 1793 the La Flavia left the harbour for Canton. The officers of this ſhip and our gentlemen had paſſed a very agreeable winter together. Their manners and behaviour were gentlemanlike throughout; nor did any of the inhabitants complain of their want of liberality.

We were now anxious for our departure alſo, but received no intelligence of the arrival of any veſſel till the latter end of July, when the Conſtantine and Helena galliot, under the command of Sturman Petuſhkoff, came into the bay from Neizſhni Kamtſhatka, whither ſhe had carried a cargo of proviſion: ſhe was now bound to Ochotſk, and put in here on purpoſe to take us on board.

We immediately embarked, took leave of our Kamtſhatka friends, ſailed the 2d of Auguſt, and arrived the 19th of the
same

fame month at Ochotſk. In this paſſage, however, we had nearly foundered. The galliot, which was ballaſted with ſand, ſprung
a leak; the pumps were clogged; and the only method was, to
bale out the water, and the ballaſt with it. However, I at length
diſcovered the leak; and Mr. Bakoff, who had been of infinite
ſervice to our Expedition in many caſes, found means to ſtop it;
but not till the water-caſks, &c. were afloat in the hold.

Application was made to the commandant for horſes; and I
went off, with the charge of the firſt party, on the 1ſt day of
September, having delivered the tribute collected at Oonalaſhka
to the chancery of the port, and obtained receipts for the ſame.

I had twelve half-ſtarved horſes, and Enſign Alexeeff and two
ſailors were with me. We had extremely bad weather, of wind,
ſnow, and rain, which retarded our progreſs very much; nor
could we poſſibly make more than 20 verſts a day. Several of
my horſes died on the road; but I received aſſiſtance from ſome
of the Yakuti, with whom I accidentally fell in as they were
returning home from Ochotſk with unladen horſes. I arrived at
Alachune with only three of the horſes that I received at Ochotſk.
Here I obtained a freſh ſupply of ſuch as were fatigued, and
hardly able to get on; and, after ſuffering inexpreſſible difficulties,
leaving my baggage behind in the woods, as alſo my companions,
in hopes of relief from my endeavours, arrived alone at Yakutſk
the 2d of October. I immediately repreſented the deplorable
ſituation in which the parties who were to follow me would, of
courſe, be placed; and that they would, probably, be loſt, if a
ſupply of horſes were not diſpatched directly for their relief,
with proviſion and other neceſſaries.

The commandant, Colonel Kozloff Ugreinin, and the captain of the diftrict, Mr. Hornoffky, exerted themfelves to the utmoft; and the fame day about 100 horfes were fent to their affiftance, and to collect my fcattered baggage, confifting of all my clothes (except thofe which I had on my back), the remains of our gold and filver medals, and other valuable articles. I was fupplied with neceffaries by the commandant (whofe clothes fitted me very well) until the arrival of my own about the middle of No-vember, toward the latter end of which month all the gentlemen, with the failors, arrived from Ochotfk. Captain Billings was the only officer of our Expedition remaining here, all the reft having embarked in the provifion veffel returning to Vircholenfk. During my fhort ftay here, I had an opportunity of feeing the Tungoofe head prince, refiding on the Aldan, near the difcharge of the river Mayo *; from whofe intelligence, in addition to the information received from Mr. Haufen and other officers of the College of Mines, I am enabled to give the following account of the Amoor, or Saghaalien.

This river takes its rife from the Kentaiham mountains, about the latitude of 49°, and longitude 110°, eaft from Greenwich; and is here called the Onon. Its direction is nearly north-eaft; and at the difcharge of the Nirza, where the city of Nortfhinfk is fituated, about the latitude 52°, it bears the name of the Shilka. This courfe it continues to the latitude $52\frac{1}{2}°$, its moft northern extremity, where the Tungoofe call it Amoor, and the Chinefe Saghaalien Ula (Black Mountain River; I prefume, from the oak forefts on the mountains hereabout, which the Chinefe call

* Alluded to in page 138. He forwarded fafely a letter directed to Mr. Saretfheff, fome-where on the coaft of Ochotfk, perhaps between the Port and the Aldama or Ud rivers.

Black-

Blackwood). From hence it is navigable in veffels of moderate fize, having received confiderable fupplies from the torrents rufhing down the eaftern and northern mountains, as alfo from a very confiderable river flowing from the fouth-weft, and called the Argoon, which difcharges itfelf into the Amoor about 180 miles eaft of Nertfhinfk. In the vicinity of thefe parts the Ruffians have feveral forts. From latitude $52\frac{1}{2}°$ to $47\frac{1}{2}°$ it flows nearly fouth-eaft, receiving in its courfe a number of rivers both eaft and weft. The Tfhukir has its fource from this fide of the fame mountains as give rife to the Olekma and Aldan * (both empty-ing themfelves into the Lena); and, flowing nearly fouth, joins with the Silempid, which flows from the vicinity of the Ud †, keeping nearly a weftern courfe into the Amoor. All thefe rivers are navigable for boats nearly to their fource.

The country is very mountainous, but the vallies and plains are fpacious and fertile. I am induced to be fo particular with regard to thefe rivers, becaufe they form a fecure retreat to fuch Yakuti and Tungoofe as are diffatisfied with their fituations about Olekma, Yakutfk, the Vilui, and Ud. Here they enjoy the pro-tection of the Chinefe, and, I am told, have built feveral ftrong places: and, as they are very numerous, they form no inconfider-able advance guard to the Chinefe frontiers.

In the year 1787, there migrated to China, from the diftricts of Olekma, Yakutfk, and the Vilui, more than 6000 Yakuti, with all their poffeffions ‡. Thefe circumftances have led me into a

---

* See page 24.   † The Ud flows into the fea of Ochotfk.

‡ This intelligence I obtained, in the houfe of the Ifpravnik Mr. Hornoffky, from Meffrs. Evers and Kyfhkin, both affeffors in the Ruffian fervice.

Thefe migrations certainly reduce the number of Tartars tributary to Ruffia.

digreffion

digreſſion from the Amoor ; and before I return to that ſubject I ſhall preſume to hazard a conjecture, that ſome future traveller may diſcover in theſe parts a nation of people unknown before, who from their mixture of Yakuti, Tungooſe, Burati *, Manzſhuri, and Chineſe, may form a new language of their own. The immenſe tracts of fertile land uninhabited and uncultivated will lead the emigrants to ſelect ſuch places as are moſt likely to produce every means of ſupport ; and they may be of great aſſiſtance to the Chineſe by cultivating of corn, &c. The low country, however, labours under the diſadvantage of being ſubject to inundations, and earthquakes are very frequent.

No rivers of any importance join the Amoor from the eaſt, except the two above mentioned. The Nonni Ula, however, a very large river, which takes its riſe about the latitude of 51°, and longitude 123°, makes a conſiderable inland circuit, and empties itſelf into the Amoor at its ſouthern extremity, about the latitude of 47½°. Another conſiderable river, the Uſuri, loſes itſelf in the Amoor nearer its eſtuary, about latitude 48½°. It riſes from the lake Hinka, and has a communication, after a ſhort day's journey by land, with the ſea of Japan. It now flows in its own channel north-weſt into the ſea of Ochotſk, about the latitude of 52½°, oppoſite the iſland Sagha-alien. This river is well ſtocked with fiſh, and its borders are covered with foreſts of oak, walnut, birch, and different ſorts of pines. The ſoil is very rich, the climate mild and healthy. The inhabitants of theſe parts of the coaſt, as alſo of Corea, and the contiguous iſlands, are not very numerous, but extremely hoſpitable and good natured, and carry on a trade with the interior for mere neceſſaries.

* See page 19.

The

The Kamtſhadals, who have viſited the ſouthern Kuril iſlands, ſpeak very favourably of the honeſty and kindneſs of the inhabitants. *I hope, however, that I ſhall ſtill be able to give a better account of theſe unknown regions hereafter, from perſonal obſervation.*

I remained in Yakutſk with Captain Billings till the 2d of January 1794, when we departed in ſledges for the city of Irkutſk, where we arrived about the middle of the ſame month, and met with all the officers of the Expedition.

We were here informed, that Lord Macartney was in China on an embaſſy from Great Britain, which led to various conjectures ; but had I received any intimation of his being expected there while I was in Kamtſhatka, or at Ochotſk, I ſhould moſt certainly have paid my perſonal reſpects to His Excellency in Pekin.

The following is all the intelligence that I could procure of Captain Billings's expedition acroſs the land of the Tſhutſki ; and for it I am indebted to the journal of one of the party.

Auguſt 13.—" At nine o'clock this morning we departed from the bay of St. Laurence, and firſt croſſed to the ſouth-ſide, when the baidars were hauled ſometimes by the Tſhutſki, and ſometimes by harneſſed dogs running along the beach. We paſſed three villages belonging to the natives, and halted at a fourth for the night. The huts were dug under ground, and covered with earth. They were of a ſquare form, with a fire-place in the middle, and four large ſtones made the hearth. They have no wood, but burn the bones of whales, pouring the oil of ſea animals

mals upon them.   Each fide of the hut contains a polog, or low tent, made of leather, to fit and fleep in.

" Our firft arrival among them did not promife much happi-nefs in their company; for, not knowing their language, we were obliged to treat with them by figns *, for fuel, water, &c. to boil our food, and pay for it immediately. Obferving our good nature, and want of power, however, they at length took a liking to the buttons on our coats, which they cut off without ceremony; they alfo ftole our fnuff-boxes; and without any hefitation paid a vifit to our portmanteaus, in hopes of finding tobacco and iron.

" The men were tall and ftout, dreffed in a neat park (refem-bling a carter's frock), made of the fkins of different animals bordered, tight pantaloons of doe-fkin, and boots of feal-fkin; the head uncovered, and the hair cut fhort.   The warrior has his legs and arms punctured, fo as to denote the number of the enemy that he has flain, and the prifoners he has taken.

" The women were alfo well made, above the middle fize, healthy in their appearance, and by no means difagreeable in their perfons.   Their drefs was of doe-fkin, with the hair on; and one garment covered their limbs and their body : this is a park, with roomy pantaloons fewn to it, and fleeves down to the wrifts. They put the legs into the opening at the neck, where it ties, as alfo below the knee.   Long boots of rein-deer's legs, with the hair on, are drawn up, and tie over the above drefs at the knee. They wear their hair parted, and in two plats, one hanging over

* I cannot conceive where Dauerkin, their interpreter, was at this time.

each

Alexander del.

Neagle Sculp.

A Tshutski Woman.

Published March 2.^d 1802, by Cadell & Davies, Strand.

Plate XIV.

Drawn by M. Alexander.

Engraved by W. Coode.

A Man in Armour with a Woman and a Child of the Tshutski.

Publish'd March 2·1802 by Cadell & Davies Strand.

each fhoulder, their arms and face being punctured very neatly *, though almoft every one differs from another in the figures. They wore necklaces, and had ftrings of beads fufpended from the ears, as alfo iron or brafs rings round the wrift.

" Auguft 14.—At eight o'clock this morning we proceeded in our boats, or baidars, entered the bay of Metfhikma, and obferved on the oppofite fhore (an ifland) a village of the fame name. We croffed this bay, and arrived at the camp of the Rein-deer Tfhut-fki, who were to be our guides acrofs the country.

" Our reception by thefe people was very ftrange. At firft they oppofed our landing; old and young, boys and girls, crying out and throwing ftones in the fea. After they had done this for fome time, the chief (who is named Imlerant) appeared, with feveral old men, and made two fires; then took our commander by the hand, and led him over one of the fires; took off his own park, and put it upon Captain Billings, who, in return, put a clean fhirt upon the chief: this exchange of drefs is confidered as a mark of friendfhip and mutual protection. The ceremony of croffing the fires was impofed on every one of us; and all our baggage, provifion, &c was alfo handed acrofs them. The chief then placed before us large pieces of boiled deers' meat extremely fat; and, to fhew our fenfe of his hofpitality, we prefented him with tobacco, beads, and needles.

" At the fetting of the fun they commenced racing and wreft-ling: it was not a race for fpeed, but running round a ring for

* The annexed ENGRAVING, taken from an original Drawing, will fhew the appearance of this their fafhionable ornament.

T t

a con-a confiderable time; and he who held out the longeft was the hero, and had the upper feat affigned him. The wreftler who overcomes all the reft is reckoned the moft favoured, as among the Yakuti.

" 15th.—Imlerant, the chief, received the following prefents to divide among the people : 2 poods of iron ; 2 poods of tobacco ; about an equal quantity of beads ; ear-rings, trinkets, and needles. Our interpreter was defired to tell them, that, in return, we hoped they would affift us with food, warm clothing, and every neceffary in their power; and, without any attempts to infult, conduct us fafely acrofs their country.

" 16th, 17th, 18th, and 19th, we had rainy weather. 20th, the herds of rein-deer were driven towards the camp, or tents, and halted on their arrival at the rivulet: upon which, two men went out with fire, and two women with fmall buckets of oil; fires were made, and the deer driven acrofs them and the rivulet to the tents ; when a round inclofure was made by the chief of each herd with the fledges of the men, and the different herds were driven into the refpective inclofures ; the women's fledges were placed between them and the fea. Fuel was now added to the fire ; the elder chief feized one of the deer, and gave it to his eldeft fon, who led it towards the fea, ftabbed it with his fpear on the left fide, and then loofened it. They pay particular attention to the manner of the deer's falling ; if on the right fide, and it dies eafy, they fuppofe that it portends good fortune, and fuccefs in their undertaking ; but if it falls on the left fide, or is convulfed, the omen is not propitious. This example was followed by the owner of every herd, each taking a handful of the blood of the ftabbed deer, which they threw firft

towards

towards the fun, then to the fea, and laftly to the mountains. When they had finifhed this ceremony, and did not purpofe kill-ing more, the women fkinned and cleaned the deer, and made fires where they had been flaughtered (every one feparate). They boiled meat, and rubbed the marrow on the faces of their idols, which they call Gir Gir (God). They have different gods,—as, of fire, of good and of evil. The idols are pieces of wood of differ-ent forms, with faces cut out, and ferve for making fires by friction.

" The next day (20th Auguft) they had a ceremonious feaft. At feven in the morning three of the flaughtered rein-deers' heads (with the horns on, and the whole fkin adhering) were placed on little benches, with two of the legs of the deer; where-upon, four of the oldeft chiefs took each a tambour, and began beating, walking gently round, and muttering fome words, raifing the voice by degrees; at laft they became clamorous, and danced. Having continued fome time, the hoft went to the fmall tents (which are covered without light) and afked thofe fitting there, " How are you ?" We could not obtain any explanation of the meaning of any part of this ceremony. Upon his opening the polog, thofe fitting within it anfwered, " Chaiyua, chai-yua, chai-yua, lewnom lewnom ;" which is, further, and further, and fur-ther,—better and better. After he had gone to all the fmall tents, they continued the ceremony, as above, for a confiderable time ; and, upon finifhing, the hoft Imlerant went to our com-mander, took him by the hand, and faid,—" We old men pro-" nounce from our obfervation, that all your undertakings will " be attended with fuccefs and good fortune ; and God has fent, " for our benefit, the Ruffians amongft us in a friendly manner, " for the firft time, to explore our fea, and reward us with li-

" berality.

" berality. God fend that we may be infeparable allies for
" ever *."

" Captain Billings immediately hung a medal round the neck
of the chief, and affured the people of the protection of Her Imperial Majefty, if their behaviour proved their fpeech to be fincere : upon which, they all bowed their heads, and cried out,
" Chayua lewnom, lewno lewnom ;" then they began dancing and
finging, men, women, and virgins, till nine o'clock in the
evening.

" The 22d, Captain Billings, Dr. Merck, Sturman Batakoff,
Draftfman Varonin, and a failor, went to the village of Metfhikma ; from which place Mr. Batakoff was fent to furvey the
bay.

" The 23d, we went on the hill to fee the winter habitations
of the ftationary Tfhutfki, who ftill refide in their tents. Here
were four earthen huts ; but three of them fo extremely filthy,
that we could not enter. One, however, we got into ; the entrance of which was formed like a watch-houfe with erect bones
of whales ; perhaps it is covered in bad weather. It was a hole
dug in the earth, eight feet fquare, and fix feet deep. The roof
confifted of whales' ribs and cheek-bones arched, nine feet high
in the centre ; the fupporters alfo were whales' bones. There
was a bench on each fide ; and the floor confifted of boards,

* Nicholai Dauerkin was interpreter. He is a native of the Tfhutfki ; was taken prifoner when young, educated in Irkutfk, and fent back to Anadirfk, with the rank of fer-
jeant, to be interpreter between the Ruffians and his own nation. This fpeech appears
to me quite in the ftile of this man himfelf, and I much doubt the truth of his interpre-
tation.

fome

fome of which lifted up for an entrance to the cellar, where they keep oils and their winter ftock of provifion. There was no fire-place, but a large difh ftood in each corner for the purpofe of burning oil *. Part of the roof was ornamented with drawings of baidars, fifhes, deer, fledges, &c. We remained here till the 25th, when we returned to the tents of our guides.

" Auguft 26.—The Tfhutfki thought proper to remove for-ward, and we proceeded to the top of a mountain 2 verfts and 25 fathoms, as meafured by a line. Here we remained all the 27th.

" On the 28th proceeded on our journey, and croffed a mountain overgrown with mofs. The next day we were joined by five tents of natives, and remained in this place till

" September 4, when we travelled one verft and a quarter, and halted all the next day.

" On the 6th we travelled by the fide of a rivulet no great diftance, and halted till the 10th. The lakes were now frozen over, and we had 7° of froft.

" September 11.—At eight this morning three rein-deer were killed with great ceremony, as a facrifice for the recovery of Owmulrat, fon to one of the chiefs, who was taken ill. As foon as they had fkinned them, they placed the fick man between the three heads, fo that his park, or garment, was over them. An

* The firft huts they entered had a fire-place in the middle.

old.

old woman whispered in the ears of the deer, and then walked round him, with lighted dried branches of juniper bushes.

" 12th, The whole of last night was passed by a sorcerer in incantations for the recovery of the sick man. This night was passed in the same manner, and the magician was paid with rein-deer.

" 13th, We remained in the same situation.

" 14th, At nine o'clock this morning the favourite dog of Awmulrat was sacrificed, being stabbed in the same manner as the rein-deer; blood from the wound was thrown three ways; the skin was taken off, the body ripped open, and the entrails examined. At noon the head was wrapped in the skin, and the sick man led round the dog, having anointed his head with blood.

" 16th, We travelled three versts and a half, and halted.

" 17th, We made one verst over a mountain, and came pretty near the bay of Metshikma again.

" 18th, Halted again. This evening, at eight o'clock, strong north lights appeared.

" 19th, We travelled close to the bay of Metshikma."

The Journal continues in the same manner, without specifying any particulars, or mentioning in what direction, until the 4th

October,

October, when they were joined by the other interpreter Kobeleff, (a Coſſac Sotnik *).

" October the 5th, Captain Billings and Kobeleff went on before with 17 ſledges loaded with the whole of the Captain's baggage. (From this time the party behind the Captain ſeem to have ſuffered materially, with regard to food, &c. ; and on the 9th the Tſhutſki ſtole the meaſuring lines.)

" 12th, Imlerant, the chief, and his wife, went on with 12 ſledges to overtake Captain Billings, to obtain ſome tobacco, &c. and to tell him to wait. We this day came to the river Ugnei, which falls into the bay of Klutchenie, and left the river on the left hand. Upon our halting for the night, the Tſhutſki compelled us to go back to the river, to ſeek on its borders ſome bruſh-wood to dreſs food. We had much ſnow and wind.

" 13th, This day we croſſed three lakes ; the firſt of 300, the ſecond 400, and the third 300 fathoms. We now ſuffered conſiderably, and could plainly perceive it to be our interpreter Dauerkin's fault ; who, when we halted for the night, aſſumed a right to prevent our getting meat ; telling us, that we ſhould not have any, becauſe we had not collected wood. Hitherto we had received frozen-meat.

" 14th, Arrived at the bay of Klutſhenie †.

* Commander of a hundred.

† I believe that the entrance of this bay forms the extreme point of Aſiatic land ſeen by Captain Cook, and called by him Cape North.

" 15th,

" 15th, We turned from the bay to the weft, after travelling its borders to fome diftance, and paffed the night by a rivulet.

" 16th, On account of bad weather, halted.

" 17th, Croffed a mountain and two rivulets, and halted by a lake.

" 18th, After croffing a mountain, we came to a confiderable river called Chainana *, but we were 70 verfts from its difcharge. This day we had nothing but raw meat allowed us, which we ate in a frozen ftate.

" The 21ft we overtook Captain Billings. He diftributed prefents of tobacco, &c. among the Tfhutfki, who readily pro-mifed to feed us well, and ufe us better; upon which he again, on the 22d, went forward with Kobeleff and the Sturman's affift-ant Gilleeff.

" 23d, Numbers of Tfhutfki paffed us, and pitched their tents at no great diftance. The chief of our party went to them; and his brother robbed us of almoft every thing that we had. How-ever, he gave us plenty of meat, boiled and raw.

" 24th, The chief returned, and we croffed a mountain.

" 25th, 26th, 27th, Halted.

* I take this to be the river that falls into the Icy Sea, a little weftward of Klutfhenie Bay.—N. B. I obferve, that this river, on the Ruffian charts, is called Amga Yan.

8

" 28th,

" 28th, We this day came to a rivulet where we obferved numbers of Tſhutſki.

" 29th, Halted.

" 30th, At nine A. M. the chief and I went on to Captain Billings, and received tobacco, beads, &c. upon which we returned to our comrades, and went to feek a feeding-place for our deer.

" 30th and 31ſt, Halted.

" November 1ſt, Halted. The reafon of halting now, I was informed, was, to kill deer for the parties going to the Kovima, which was 250 * verſts diſtant from this place.

" 2d, I was fent forward, under the charge of the ſiſter of the chief, with two ſledges, and went about three verſts, when we halted, and were afterwards joined by the reſt.

" 3d, Halted. The 4th travelled, I ſuppoſe, about 16 verſts.

" 5th, Came to a large river, about which ſeveral parties of Tſhutſki were travelling. We halted near a conſiderable body of them, having travelled, I believe, about 20 verſts."

The journal goes no farther; and I had no opportunity of procuring any explanation; but I believe this is the place where one attempt was made to maſſacre the travellers; in all probability,

---

* Perhaps the river Angarka is meant inſtead of the Kovima.

through

through the perfuafions of Dauerkin *; but the other interpreter, Kobeleff, fufpecting their defigns from their motions and their converfation, acquainted Captain Billings with their intention, and immediately called the chiefs, told them that he knew what they were about, and faid, " We are all ready to die; but remember, our bones will be found, and raifed by the Ruffians, although you burn them to afhes." Upon hearing this fpeech of Kobeleff's, they confulted together, and continued their journey, promifing not to kill them.

The other remarks made by the writer of the journal that I have tranflated, are fuch as I have already taken notice of; except that the large baidars of the ftationary Tfhutfki are all made of one fize, and upon one plan, covered with the hide of the walrofs, and rowed with eight paddles. Befide thefe, they have fuch covered ones as the Aleutan iflanders have, with one and two feats, but much heavier. The wandering tribes confider themfelves more independent than the ftationary, and will not allow their wives or flaves to have any intercourfe with ftrangers; while the ftationary tribes admit of this without any kind of hefitation, particularly with their flaves. Thefe, however, are treated very differently from fuch as are free; and it fometimes happens, that when the latter are not fatisfied with their fituation, they leave one man and go to another. I cannot give any further information refpecting thefe people.

At Irkutfk every poffible difpatch was ufed to finifh the part of our bufinefs which depended upon that government; and at

---

* It is my opinion, that this man, who was of a fullen, jealous, and revengeful difpofition, found himfelf hurt by the confidence which the travellers placed in Kobeleff. He had entertained thoughts of rewarding his Tfhutfki friend, and appearing a man of confequence among his countrymen; in which, however, he was completely difappointed.

the

the latter end of the month of January we set out for St. Peterfburg by the fame route which we had taken hither. The Siberian
inhabitants appeared rather more fhy than they were in 1786;
perhaps owing to the feafon, it being lent. They alfo complained, that their intercourfe with the Mongals was not fo open
as formerly, the latter having retired to the Chinefe frontiers.
The Tartar women about Tara were preparing nettles, and fpin-
ning thread from the fibres; the linen made of which was fine
and good, apparently equal to that made of flax. I have already
mentioned the neat carpets then made by thefe induftrious people.

I was furprifed at the appearance of detached families of Gipfies
throughout the government of Tobolfk; and upon inquiry I
learned, that feveral roving companies of thefe people had ftrolled
into the city of Tobolfk. The Governor thought of eftablifhing
a colony of them; but they were too cunning for the fimple
Siberian peafant; which induced him to feparate each family.
He placed them on the footing of the peafants, and allotted a portion of land for cultivation, with a view of making them ufeful to
fociety. They, however, rejeĉt houfes even in this fevere climate,
and dwell in open tents or fheds; nor can they be brought to any
regular courfe of induftry; but they watch every traveller, and
pretend to explain the myfteries of futurity, by palmiftry or phy-
fiognomy. The peafant dreads their power, and from motives
of fear contributes to their fupport, left they fhould fpoil his
cattle and horfes. It is faid, that they are very fkilful farriers
and cowleeches.

I obferved the whole way back a confiderable diminution of
trees; and in the vicinity of Ekaterineburg, and all the iron ma-
nufaĉtories, where the road led through forefts that appeared on

4   either

either fide impenetrable, we now croffed plains where hardly a tree was left ftanding : this was the cafe nearly contiguous to the new made towns, and on the borders of the navigable rivers. Immenfe quantities of timber are floated down the Volga into the Cafpian and the Sea of Azof, for private and public ufes, as alfo for further exportation ; and, as the generality of buildings throughout the interior of Ruffia and Siberia are conftructed of timber, fires frequently confume whole towns and villages; nor have they any other fuel than wood; for, notwithftanding pit-coal is in many places in great abundance, it is never made ufe of. It would be greatly to the advantage of the country to en-force the building with more folid materials, and to encourage the ufe of coal for firing ; particularly for the different works that confume much fuel *.

I arrived in St. Peterfburg on the 10th March 1794, fo very much afflicted with the rheumatifm, from a cold caught at Ir-kutfk, that in regard to action I was reduced to the helplefs fituation of an infant. The kind attendance, however, of Doctor Rogers, and the friendly affiftance of the Britifh merchants in that city, who are fo eminently diftinguifhed for their unbounded hof-pitality, alleviated every pain, leffened every difficulty, and pre-vented the miferies of penury from being added to my mif-fortunes.

* Since my return from the Expedition, I have been conftantly travelling about the fouthern borders of Ruffia, and have feen pieces of pit coal in the Oka and Volga rivers ; and all the country between the Dor and the Black Sea is replete with coal. Regular pits are funk about 200 verfts north of the fea of Azof, by Englifhmen in the employment of Mr. Gafcoigne, for the ufe of the Black Sea fleet, and of an iron foundery lately built near the river Donets.

APPENDIX.

No. I.

VOCABULARY

OF THE

YUKAGIR, YAKUT, AND TUNGOOSE (OR LAMUT) LANGUAGES.

N. B. Ch muft always be pronounced like the German *ch.*—*I* is always fhort.

Englifh.	Yukagir.	Yakut.	Tungoofe.
God	Chail	Tangra	Gheooki
Father	Etchèa	Agam	Amai
Mother	Amea	Iya	Eni
Son	Antoo	Oal	
Daughter	Marhloo	Keefim	Ghoorkan
Brother	Tfhátfha	Oobagim	Akan
Sifter	Pawa	Agafim	Ekin
Hufband	Yádoo	Erim	Edee
Wife	Alwáley	Yaghtarim	Akee
Maiden	Váiendéndi, Marchet	Keefa	Choorkan
Boy	Luhundæ	Ogo	
Child	Lukoolu (in arms)	Kutu ogo, little boy	Kootian
	Uwá (beginning to walk)		
Man	Toromma	Kiffœ	Bey
Men	Toromma	Kiffœlar	
Head	Iok	Bafs	Del
Face	Neatfha *	Sirai	Itti
Nofe	Iongul	Mooron	Ogot

* Neatfhaga, the fkin of any animal.

[A]

Noftrils

English.	Yukagir.	Yakut.	Tungoose.
Noſtrils	Iongundangil	Tani	K-elon
Eye	Angzſha	Kaſak	
Eyebrow	Angzſhabugúelbi, alſo eyelaſhes.	Chas	Karamta
Ear	Oonómma	Kugach	Korot
Forehead	I-óanguitſhel	Süis	Omkat
Hair	Manalláe	Aſſim	Nioorit
Cheeks	Moonéndzſhi	Singak	Antſhin
Mouth	Angá	Aiyach	Amga
Throat	Tónmúl (hunger tonmulla)	Kaima	Belga
Lips	Anghenmóoga	Ooas	
Teeth	Tòdy	Tees	Itſhi
Tongue	Onnór	Till	Enga
Beard	Angénbugüelbi	Buitik	Tſhurkan
Neck	Jomüel	Moinung	Mixon
Shoulder	Nungénmoogá	Saning	Mir
Elbow	Itſhe-endamey	Tongonock	Etſhen
Arm	Núngean	Illi	
Hand	Nugán		Gal
Fingers	Pe-enditſſia		Kabr
Finger-nails	Onzſhil	Tingrach	Oſta
Breaſt	Mélud	Tueſs	
Belly	Líeril	Oſſogo	Oor
Back	Jewóghá	Sies, Kochſui	Neri
Feet	Noel	Attach	Boodel
Knee	Tſhorkel	Tuelgeſſo	
Heart	Tſhóoenzſha	Surach	Mewan
Stomach	Niméngſhinzſhá	Mungra	
Blood	Liòpkul	Ghan	Soogial
Milk	I witſhi	Ee-ut	Ookiooln
Skin, hide	Char, alſo clouds	Tiri	Nandra
Meat	Tſhul, alſo body	Et	Oolra
Bones	Amún	Umok	Ipree
Hearing	Mōēdik, heard, alſo felt	Iſſit	Iſni
Seeing	Umat, ſeen	Anar, Koer	Igoorun
Taſting	Tſhangitſh, taſte	Amtan	Amtam
Smelling	Lemlemoodel, ſmell	Sitta, Seligan	Moyeni
Felt	Moedik	Iſtebin	
Voice	Orni	Koemoya	Delgan

7.

Talking

English.	Yukagir.	Yakut.	Tungoose.
Talking	Aniak	Ittare, speak, Kapsir	
Name	Nevĕ	Aatta	Gerbin
Scream, crying out	Orinak	Sangarda	Irkan
Outcry	Orool	Kittanar	Mogandra
Noise	Mungzsha	Yedeimeng	Ooldan
Crying	I-véllek	Ittir	
Tears	Angzhanondz shi		
Laughter	No-ok, laughing	Koïller	
Sneezing	Tshangnūi	Ittereer	
Scratching	Pandalitsh		
Trembling	Lirkúndzshi	Tittirir	
Singing	Jagtak	Toy	Ikan
Sighing	Ningelamoditsh		
Whistling	Tshundzsha	Issir	
Lie down	Kondāk	Sit	
Go	Ingherghodak	Bar	Choorli
Stand	Onghak	Tur	
To Sleep	Iūnzshul	Tui	Ookladai
Dream, sleeping	I-unzshuk	Oomkella	Ooklean
Jumping	Moēnmōēnga	Ekerek	
Holding	Ma-ik	Tut	
Running	Tshuenzshi	Suir	
Dancing	Longdok	Inkullæ	
Love	Anoorak	Tapta	Googemon
Lover	Anooroh	Taptasabit	
Glad	I-ak	Yarabin	
Joy	I-ai	Yurdim	
Sorrow	Artshetshúnzsha	Sanangatim	
Pain	Joatsh	Irridim	Eyen
Trouble	Ankorfy	Irridenim	Choonatsh
Work	Ooil	Illulatim	Goorgalden
Laziness	Alangnae	Surugaldzshitim	Ban
I	Matak	Min	Bee
Thou	Tat	En	Boo
He	Tundal	Ginne	Nongenatshe
We	Mitek	Buissiga	Nonganoobe
Ye	Titlak	Issige	Ellia
Eating	Langdal, ate	Assibin	
Thou eatest	Tatlak	Assa	
To drink	Ondzshok	Issiem	Koldakoo
To feed	Sagetak	Assiapin	
			Taking

English.	Yukagir.	Yakut.	Tungoose.
Taking	Mendzſhit	Illiem	
Carrying	Moream	Ildzſhi	Ghenoom
Throwing	Potſhitſhik	Brach	
Giving	Keick	Beer	Omool
		Give me aghal	
To cut	Tſhok	Buis	Minadai
Hiding	Angítak	Kiſtya	
Beating	Kogdak	Sienem	Madia
Strength	Tonboy	Kuiſtak	Egooi
Birth	Oo-inge	Terretpuit	Ekzſhecan
Race of people	Ommo	Omung	Beyil
Marriage	Torroi	Kurum	Awlan
Widow	Poóndalvólle	Erimſoch Yagtar	
Life	Liak	Olloruput	Inni
Body	Tſhul	Ettim	
Spirit or ſoul	Liéuſha	Tina	
Death	Amda	Elbuta	Kokan
Age	Ligai	Kerdzſhagas	Sagdi
Youth	Andelgoin	Edder	Noolſóolktſhan
Large or great	Tſhomoi	Oolachan	Ekzſham
Small	Lukun	Kutſhugai	Nukiſhookan
High	Pudanniai	Irduk	Gooda
Low	Ledemnie	Namtſhiltſhak	Netkookak
Cold	Pondzſhetſh	Timnee	Iguin
Warm	Pugatſh	Ettegas	
Hot	Pugatſh	Itti	Ghochſin
Health	Tauritſh (good) well	Ellérbuin, Ittugai	Abgar
Malice	Erritſh	Kuttir	Booktſhalran
Stupid	Evĕntſh	Mennek	
Wife	Onmanneig	Kerſie	
Agreeable	Naintallitſh	Ittugai	Ariooldooln
Sharp	Natſhennee	Sitti	
Round	Pomne	T'ungruk	
Circle	Pomdzſhólené	Tungrutſhu	
Ball	Loatſha		Mewreat
Light (weight)	Arrángiā		
Heavy	Ningoin	Tſhiptſhik	Aimkoon
Strong	Addi	Oorachan	
Weak	Nóndri	Kittanach	
Tight	Iklon	Meltoch	
Thin	Ke-ivey		
		Sinnegas	
			Thick

English.	Yukagir.	Yakut.	Tungoose.
Thick	Inglon	Soan	Derom
Broad	Kanbunnoi	Ketil	
Quick	Omduk	Turgan	Oomufhat
Gentle	Anindzfha	Argooi	Etnioo
White	Po-innei	Irungk	Geltadi
Black	Aimáivi	Chara	
Red	Kelenni	Kafil	Koolani
Green	Tfhakolonni, alfo yellow	Keoch	Tfhulban
Blue	Lubanzfhanni	Keochtinoo	
Sun	Jelónfha	Kuin	Nultian
Moon	Kininfha	Ooi	Begh
Stars	Lerungundfhia	Solus	Ofikat
Sky	Kundfhu	Chaltan	Gioolbka
Fog	Tarrel	Kudon	
Clouds	Char	Bullit	
Sun's ray	Jelondfhendigia		Elganee
Wind	Illejénnie	Tyil	
Blowing	Pookindfhi	Kotutar	
Whirlwind	Jadondajendelaia		Ghuee
Storm	Tfhemondilaia		Khuga
Steam	Leutfhénni (mift)		Okfin
Rain	Tiba	Samir	Oodan
Thaw	Nunbur	Chafing	
Hail	Jarchandiva	Tollon	Bota
Thunder	I-endu	Eting	Afhdoo
Lightning	Borongille	Tfhagilgan	Tapkitan
Snow	Pukoélli	Char	Imandra
Ice	Iárka	Boos	Bookus
Fire	Lotfhel	O-at	
Light	Pondfhirka, alfo day	Sirdik	
Shadow	Ivi	Kuluk	
Dark	Emmitfh	Kharanga	
Day	Pondfhirka	Kuin	Ining
Night	Emmel	Tuin	Golban
Morning	Unhaiel	Erdee	Tek
Evening	Poinjuletfk	Ke-effe	Moorak
Eaft	Jelongédukfhimba, fun rife	Kuintachferra	
Weft	Aivinda	Onga	
North	Ledinda	Illin	
			South

English.	Yukagir.	Yakut.	Tungoose.
South	Pondzſhirka putel (mid-day)	Sogree	
Summer	Puga	Sacin, ſoyin	Anganal
Winter	Zſhendſha	Kiſun	
Autumn	Nada	Kuiſſin, kuiſan	
Spring	Pora	Saas	
Year	Nejunmolgul	Sil	Angan
Time	Indada	Tſhitſhimtſhee	
Earth	Levje	Sirr	Tor
Water	Ondzſhi	Oo	Moo
Sea	Tſhobul	Baighal	Nam
Lake	Jalgyl	Koel	
River	Onnong	Yrris, yrrach	Okat
Rivulet	Onnongi, nalitſha		Okatſhan
Waves	Moinchaija	Duogun	Bialga
Iſland	Ommul	Arre	
Sand	Nongha	Kumach	Ooneang
Clay	Glina	Boar	Telba
Duſt	Pogintſhi	Boar kotta	Ch-engelren
Dirt	Kundun	Barri	
(Hill) mountain	Pēa	Seer	
Shore	Ighil	Kittæ	Ch-oolin
Depth	Tſháginmon	Dirring	Choonta
Height	Pudenmai	Irduk	Oſkiaſookun
Breadth	Kanbúnnai	Ketit	Demzſha
Length	Tſhitnai	Uſtata	Ghonamin
Hole	Kondzſha	Chaiagas	Changar
Grave, or ditch	Inghis	Een	Chooneram
Rock	Pea, alſo ſtone, mountain	Taas	Dzſhool
Iron	Lundal	Timir	
Salt	Logodúntſhinu, and Nimedzſhindſha	Tus	Tak
Weeds	Oolega, alſo graſs	Keoch	Orat
Tree	Tſhall	Maſs	
A wood	Jungul	Tya	Kenita
Root	Larkul	Turdæ	Kobkan
Stump	Koikél	Tſhongatſhok	
Bark	Tſhangar	Chalterik	Oorta
Branch	Tſhilga	Buſuk	Gar
Leaf	Paldſhitſha	Seberdak	Ebdernia

Flower

English.	Yukagir.	Yakut.	Tungoose.
Flower	Poelri	Dzſhuſin	
Berry	Leviéndi	Otton	
Field (plain)	Pondſhórkoni	Chodu ſaſir	
Beaſt	Talau	Koeil	Boyun
Fiſh	Annil	Balyk	Olra
Worms	Kalnindſha	Iyene	Ogil
Frog	Alundala	Baga	
Fly	Nilendoma	Zachſirga	
Ant	Jojakondzſha	Kmirdagas	
Spider	Managadaibi	Oguigos	
Argali, ſheep (wild)	Monoghá		Ooyamkan
Dog	Tabaha	It	Nin
Mouſe	Tſhalbŏe	Kutuyak	Tſhalooktſhan
Gooſe	Landzſha		Erbatſh
Duck	Ondzſhinonda, wa-ter-bird		Neki
Feathers	Pugelbi, or hairs of beaſts	Charungatſhæ	Detle
Eggs	Nontondaul	Simmit	Oomta
Neſt	Awoot	Oyo and Oyetto	
Shepherd	Itſhel	Maniſit	
Hut	Numa	Balagan	Dzſho
Door	Anbandangel	Dzſhel	Oorka
Hearth	Eviér	Kolumtan	Nerka
Floor (earth)	Liebe	Sir	
Hatchet	Noomundzſhi	Sugai	Tabor
Knife	Tſhagoia	Buſak and Buhak	
Boat	Aktſhel	Bat	
Carrying	Elléyik	Teyachpit	
Building	Aak	Ongroch	
Cloaths	Mããjil	Tangas	
Food	Lagul	Aas	
Raw	Onje	Sikai	
Dreſſed	Pánduk	Buſar	
Thief	Olonunga	Orſach	Dzſhioormin
War	Neretſhángaté, and Chimdzſhingi	Serri	Chooniat
Quarrel	Illedangi	Jegu Yegu	Dzſhargamat
Fighting	Chimdzſhingi	Ellerſy	Kooſikatſhin
Spear	Tſhovina	Innie	
Guard	Itſhell	Kettebil	Goodatſh
			Diſtreſs

English.	Yukagir.	Yakut.	Tungoosо.
Distress	Oo-ilgaitsh	Aldzsharkoi	Urgadoo
Victory	Aldzshitsh	Samnardabit	Dabdaran
Friend	Aghéma	Doghor	
Enemy	Irritshundzshitoroma evil disposed man	Estiagun	
Servant	Poā	Kolutang	
Chief	Alnindsha	Toyon	
Writing	Tshorillatsho	Surrui	
Numbers	Tshungum	Achsi	
One	Irken	Bir	Oomun
Two	Antachlon	Ikke	Dzshur
Three	Iālon	Ews	Elan
Four	Iēlahlon	Tirt	Digon
Five	Enganlon	Bes	Tongon
Six	Malghialon	Alta	Kilkok
Seven	Purchion	Setti	Etgatanok
Eight	Malgialachlon	Ogos	Tshokotenok
Nine	Chuniirki-ellendz-shien	Tagos	Tshakatanok
Ten	Kuni-ella	On	Tshomkotak
Twenty	Attachongoniella	Surbey	Katshat-kotako
Beginning	Kudalaraga	Manna gitta	
Ending	Itshagi	Kotshu gitta	
Yes	Tat	Ak, ah, eh	Ya
No	Oiley	Soch	Atcha
Now	Indzshi	Billigin	Dzshoole
Before	Angnuma	Oonut	Essemek
After	Indada	Chodzshit	S-si
Here	Tia	Manna	
There	Talay	Onno	Tala
Yesterday	Nengandshé	Beghassæ	
To-day	Pondzshirkoma	Begun	
To-morrow	Ongóiē	Sarsin	
Look	Tindij	Boo	Er
How	Kondamīel	Chaitak	On
Where	Kolae	Kanna	Illey
When	Chánnin	Kassan	Ok
What	Liōmlentak	Tugui	Ek
Who	Chinetta	Kiminen	Ni
With what	Lumun	Tugonon	Etsh
Under	Tangmuinal	Allara	Ergudalin
Upon	Pudendago	Eussæ	Widalin.

VOCABU-

No. II.

VOCABULARY

OF THE

LANGUAGES OF KAMTSHATKA, THE ALEUTAN ISLANDS, AND OF KADIAK.

Englifh.	Kamtfhatka.	Aleutan.	Kadiak.
God	Newfteachtfhitfh	Aghuguch	
Father	If-ch	Athan	Ataga
Mother	Naz-ch	Anaan	Anaga
Son	Pa-atfh	L'laan	Avagatoga
Daughter	Sooguing	Afhkin	Panigoga
Brother	K-tfhidzfhi	Choyotha	Ooyitaga
Sifter	Kof-choo	Angeen	Alkaga
Hufband	Skoch	Oogeen	Ooinga
Wife	Squa-aw	Ai-yagar	Nooliga
Maiden	Ch-tfhitfhoo	Oogeghilikin	Aghanok
Boy	Pahatfh	Anekthok	Tanoghak
Child	Pahatfhitfh	Oofkulik	Tfhagaloi
Man	Ufkaams	Toioch	Sewk
People	Quafkoo, Ufkaamfit		Amalachtel-fewt
Head	T-choofa	Kamgha	Angloon
Face	Qua-agh	Soghimagin	China
Nofe	Kaankang	Anghofin	Knak
Noftrils	Kaang'a	Guakik	Padzfheeguak
Eye	Nanit	Thack	Ingelak
Brows	Tittan	Kamtic	Kubloot
Lafhes	Tfhuanit	Kochfaki	Chamagate
Ear	E-ew, E-ewt	Tottufak	Tfhewdek
Forehead	Tfhilgua	Tanneek	Tfhoo-uga
Hair	Koobit	Emley	Neweyet
Cheeks	P-phaad	Ooluga	Ooluak

[B]

Mouth

English.	Kamtſhatka.	Aleutan.	Kadik
Mouth	Kuz-ha	Aghilga	Kannak
Throat	Quiqua	Stſhoka	Yoamun
Lips	K'kovan	Kotſhoon	Keh-look
Teeth	Kuppet	Aghalun	Choodit
Tongue	Nutſhel	Aghnak	Ooloo
Beard	K'ko-ookat	Inglaak	Oongai
Neck	Hitle	Oo-iyo	Ooyagut
Shoulder	Tanutar	Kanglee	Tooik
Hand, arm	Settoo	Tſha	Ai-igit
Fingers	P-koida	Atchon	Shovgait
Nails	Ko-uda	Chagelgin	Stoot
Breaſt	Ingátáh	Simzſhin	Tſhekiaiat
Belly	K-ſoch	Kilma	Akſ-yek
Back	Altſhoo	Tſhundra	Koak
Foot	Tſh-quatſhoo	Kita	Itiat
Heart	Nókguek	Kanogh	Kanok
Blood	Méſſon	Aamyek	Kaiook
Milk	Nókkol	Makthamtanga	Mook
Skin	Koo-ogh	Katſhka	Amek
Meat, or fleſh	T'háltal	Oolow	Kamok
Bone	T'hamtſhoo	Kaghna	Nenoat
Hearing		Toltakoning	Nitaa
Seeing	Kwatſhquikotſh	Okokthakon	Tangha
Taſte	Sa-ooſen	Katha	
Smell	Skeſich	Igutſha	Tſinago
Feeling		Sitchatſhada	
Talking	Kahalkan	Toonootha	Neogtok
Name	Hágaach	Aſſia	Atcha
Noiſe	Ki-ichkich	Imatſha	Tulchoo
Crying	Kooga-atſch	Kaighalik	Keagóok
Laughing	Kaſſoogaatſh	Aloktalik	Ingliachtoak
Singing	K-tſheemgutſh	Anogatha	Attoa
Groaning	Attaſich		Knaook
Lie down	Kanhilkitſch	Thirkaigada	Inaghna
Stand	Kaſichtſhitſh	Ankakthalik	Nanaghna
Go	Kowiſitſh	Itſha	Achook
Come	Koquaſitſh	Agatha	Taieechook
Running	Kaſchiatſh	Angaiakatha	Kemaktoak
Dancing	K-hogdaſitſh	Achatha	Chelagtoak
Love		Kingochthaka	Kanogata
Joy	Kabaſik	Iglai	
			Grief

English.	Kamtſhatka.	Aleutan.	Kadiak.
Grief	Quadaſis	Alchologothik	Anchagooh
Pain		Nanalik	
Labour	Khaſus	Aguaſutha	
Lazy		Sochtalik	Kſatachtook
I	Komma	Keen	Chooi
Thou	Kiz	Ingaan	Chlput
Eat		Kaängen	Pittooaga
Drink		Taangatha	Taanagok
Take	Kommogata	Sulagna	Teooka
Strike	Takſu	Toogalik	Tſhuzſhutekew
Throw	Tſ-chluk	Ignekan	Ch-kakoo
Strength	Takaſna	Malalookan	Oonachkiktook
Marriage	Ktſhiza	Aſikſagathan	
Widow	Sooſoo	Oſchalik	
Life	Kaitaſitſh	Anghogikoo	
High	Koo-ung	Kaiakok	Kunachtook
Low	Iſ-ung	Kaielakon	Chkidok
Body		Ooluk	Kainga
Death		Aſchalik	Tokook
Big		Taangoellik	Angoch
Little		Aangonolokn	Meyoch
Cold		Kinganalik	Potſnatok
Heat		Tſhingleſelik	
Hot	Kikak	Akivachſelik	Nogtoak
Good		Tſhizſhelik	Azigtoak
Bad	Adkang	Matchizſhelikan	Kabigwaſkak
Wife		Siniktulik	Ooſewitok
Stupid		Anghagelikin	Naloo-oomok
Light, not heavy		Igthaghatok	Ogichtoak
Hard	Kittanua	Tungachſich	T-choak
Thick	Homono	Anatulik	Leegoak
Thin		Anatalokon	Amedoak
Broad	Kutenoo	Kaghtoolik	Kangatoak
Quick		Angaiak	Tſhukaladn
White	Attagho	Komakuk	Katchtoak
Black		Kaktſhiklúli	Toonongoak
Red	Tſhaang	Aluthak	Cowigtoak
Green	Nochſonne	Tſhidthgaiak	} Tſhunagtoak
Blue		Kaktſhugthuk	
Sun	Qua-atſh	Akathak	Madzſhak
Moon		'l'oogithak	Eghaloak

Star

English.	Kamtſhatka.	Aleutan.	Kadiak.
Star		Sthak	Aghia
Heaven, ſky	Kochan	Inkak	Killak
Miſt, clouds	Miſſahan	Inkamaguk	Amaigalok
Wind		Mathuk	Kaiyaik
Rain	Tſhukutſhoo	Tſhiotakik	Kidak
Thunder		Shulukſhik	
Snow		Kaneek	Anneg
Ice		K'thak	Tſhigoo
Fire	Pangitſh	Kignak	Knok
Light		Anghalk	Tangeechſtok
Dark	Doohſae		Tamleſtok
Morning		Kilak	Oonamin
Evening		Angalikingan	Akaatoch
Night	Kolkwa	Amgik	
Day		Anghalik	
Eaſt		Kayathak	Oonulak
Weſt		Tſhedulik	Tchlanik
North		Kighaithok	Oaiſiak
South		Namatha	Ooagtok
Spring		Kanikinga	Ognakak
Summer		Seahkothok	Keegtok
Winter		Kanagh	Ookſogtok
Autumn		Seahkothoking	Ookſaghtok
Year	T-chaſioo	* Kanaghinalik	
Earth	Symt	Tſhekak	Noona
Water	Ee-ee, or i-i	Taangak	Taangak
Sea	Ningl	Alaghok	Imak
Waves	Kiaha	Thuk	
Iſland	Samatſh	Taangik	
Sand		Tſhooguk	Kightak
Clay		Tſhikthuk	Kaguyœ
Mountain	Aal	Ghaiok	Kogoo
Shore	Hite-ſhoo	Atſhida	Ingat
Hole		Tſhanok	Tſhaak
Ditch		Tſhagak	Piaganok
Copper		Kanuyak	Lagut
Iron	Quatſhoo	Komlegu	Kaunooyat
Salt	Pepum	Attagook	Tſhauik
Animal			Tagaiook
Fiſh	Etſhoo		Oongooalihat
			Ekachlewt

* Winter's approach.

Worm

English.	Kamtfhatka.	Aleutan.	Kadiak.
Worm	Chubbut	Lokaiak	Kobellewt
Fly	Quamoftfh	Oolinik	Kwielewt
Plants	Sezda		Obovit
Tree			Kobogak
Bear	Kafa	Tanguak	Tagookat
Dog	Kofsa	Uikuk	Pewatit
Fox	Tfhafalhai	Okotfhing	Kaffiak
Goofe	Kfoais	Llak	Nachklaiit
Duck	Alfhingufh	Tfhakutfhadok	Sakoligak
Egg	N-gach	Shamlok	Mannik
Neft	I-i-itfh	Tfhungangen	Oongolut
Hut	Kifut	Ooladok	Tfheklewit
Door	Nutfhoo		Amik
Hatchet	Kvafqua	Anigafhip	Anigin
Knife	Watfhoo	Omgazfhizfhik	Tfhangielk
Kettle	Kukua	Afhok	Afok
Raw	Sohang	Kangakok	Ai-ce-patnok
To boil	Koquafoch	Oonatha	Kannegtok
War		Saigik	
Thief		Tfhkalkan	Teglunachtoch
Quarrel	Situngfh	Amaghilik	Aieevoak
Spear	Quaquanutfh	Kadmagufhak	Pannah
Friend	Kallal	Kinoghtaka	Tfhuaga
Enemy		Kinoghtatkakan	Tfhuugunitaga
Warrior		Kallochalik	Tfhekchuyak
Mafter	Annanum	Tokok	Anayakak
Servant	Tfheguatfh	Talha	
Yes	La	Aang	Aang
No	Ifki	Mafelikan	Pedok
Now	Daangoo	Angaiak	Chvenigpak
Before	Koomat	Angaiaktafatha	Itfi-o-ak
After	Namfako	Amoomotaflikan	Ettakoo
Nigh	Do-ok	Wagagnaghikok	
Far off	Nifch	Amathalik	
Here	Noot	Wallignakuk	
There	Onga	Amatkulikuk	
Yefterday	Aati	Kéllagon	Koagh
This day	Daangoo	Vonangalik	Gaunegpek
To-morrow	Bokuan	Ilkellagon	Oonagoo
Where	Natfha	Channa	Nai-ee-ma
How	Nochkuis	Alkólli	
When			

English.	Kamtſhatka.	Aleutan.	Kadiak.
When	Itta	Iyem	Kakoo
What	Nokai	Alkoſigtatima	
Raven		Kalkagiak	Kalnak
Eagle		Tinglak	Koomogik
Bow		Saidegich	Kitſiak
Arrow		Agidak	Chook
Darts		Agalgch	Pannah
One	Kemmis	Attakon	Alcheluk
Two	Nittanoo	Alluk	Malogh
Three	Tſhuſquat	Kankoon	Pingaien
Four	Tſhaſcha	Shitſhin	Stamen
Five	Koomdas	Tſhang	Taliman
Six	Kilkoas	Attoon	Agovinligin
Seven	Ittachtenu	Olung	Malchongun
Eight	Tſhoktenu	Kamtſhing	Inglulgin
Nine	Tſhaktanak	Sitching	Kollemgaien
Ten	Komtook	Haſuk	Kollen
Eleven	Diſukſin	Attakathamatkich	Alchtoch
Twenty	Kaſkumtuker	Algithematick	Suenak
Thirty	Tſhukumtuker	Kankuthematik	Pingaienkollen.

The Vocabulary of the Tungoofe or Lamut Language I obtained from Mr. Koch the Commandant of Ochotſk, who ſucceeded Lieutenant-Colonel Koyloff Ugreinin; the reſt were all taken by myſelf on the ſpot with great care and attention; and having had frequent opportunities to prove them with different natives, I can pronounce them correct.— There are many words in the Language of Kamtſhatka that I was not able to pronounce, and could not of courſe attempt to convey any idea of their ſound, which is the cauſe of ſo many blanks.

No. III.

No. III.

A

L I S T

OF THE

DIFFERENT STAGES FROM ST. PETERSBURGH;

Specifying the number of verfts according to which I paid for horfes; the time of arriving and departing from each ftage, beginning each day at noon, and reckoning twenty-four hours to the day.

From St. Peterfburg	Verfts.	From Zimnagorka	Verfts.
To Tzarfco Zelo	22	To Yadrova	20
Izfhora	13	Zotiloffki	36
Tofni	23	Vifhne Volotfkoi	36
Lubani	26	Vydrapufk	33
Tfhudova	32	Torzfhok	38
Spafkoi Polifti	24	Mednoi	33
Berezovoi	24	Tweer	30
Novogorod	22	Gorodki	28
Bronitfa	35	Davidova	26
Zaitfova	27	Pefki	31
Kreftfi	31	Tfhornoi	23
Ezfhelbitfi	38	Mofco	28
Zimnagorka	23		

Places.	Verfts.	Date, 1785.	Time of Arrival.		Departure.	
From Mofco - -		Dec. 15.			4.	
Novaja - -	24		7.		8.	
Bunkova - -	34		10.	50	12.	
Kerfhatfhi - -	34		16.		17.	
Petufhki - -	26		20.		21.	
Undal - -	27	16.	24.		1.	
Valadimer -	28		6.	30	7.	30
Tfhudogda -	39		13.	50	14.	30
Mofhok - -	30		19.	15	20.	
Darfhevo -	27		23.		23.	30
Muroma - - -	30	17.	2.		3.	30
Monakova -	25		7.	30	8.	30
Pogoft -	29		15.		18.	
Pavlova -			21.			
		18.			7.	
Lafhkova						
Neizfhnei Novogorod -	33		19.		21.	30
Befvodnoi -	34	19.	4.		9.	30
Tatnits -	29		15.		19.	30
Oftafhick - -	27		22.	45	23.	40
Ofinka - -	32	20.	2.		4.	
Yemangafh -	32		8.		9.	10
Scartog - -	32		11.	30	13.	40
Atchkarene - -	22		16.		16.	35
Tfhebakfar - -	30		21.	10	23.	
Kofhki - - -	26	21.	3.		9.	
Ganafh - -	30		16.		16.	40
Vefovigh -	31		21.	25	24.	10
Kazan - -	24	22.	4.		4.	30
	30		8.	30		

Places.	Verfts.	Houfes and Churches †.	Date, 1786.	Time of Arrival.	Departure.
From Kazan			Jan. 10.		1.
To Beruli, village -	30	70		6.	6. 30
Arfk, city - -	26	100		10. 15	10. 45
		† 2			
Karadvan, village -	28	20		18. 45	20.
Jangulov - -	18	100		22. 40	24.
Gunbar - -	29	49	11.	5.	7. 50
Teremefe - -	42	40		16. 45	17.
Vazfhintech-Kakfe -	26	17		20. 30	21. 10
Sumfae - - -	20	36		23. 15	24.
Ubarie - -	14	25	12.	2.	2. 15
Kulmetfat - -	29	25		5. 30	6. 30
Zaitfi - - -	38	20		10. 25	11.
Igra - -	37	25		17.	17. 20
Bolfhoi-Purga -	32	50		21.	22.
Debeffa, village -	25	90	13.	1.	2.
		† 1			
Sofnova - -	52	200		10.	10. 30
		† 1			
Dubroffky -	25	35		13.	13. 20
Ochanoi, city -	26	60		17.	17. 45
		† 1			
Poldenoi, village -	16	20		20. 30	21. 15
Kultaiva	28	40	14.	1.	1. 15
		† 1			
Koianova -	25	80		8. 30	10.
Krilufova -	37	100		19. 15	19. 45
		† 1			
Kungur, city -	20			22. 15.	
			15.		21. 30
Stretenfkoi, village	26	80	16.	2. 45	4.
Zolotuouftoffky	19	330		7. 40.	11.
		† 1			
Bukovi	16	60		13.	13. 30
Atchinfky Krepoft -	20	100		17.	17. 30
		† 1			
Bifirfki do. -	20	100		21. 15	23.
		† 1			
Klenoffky do.	22	70	17.	3. 15	5.
		† 1			

[c]

APPENDIX. No. III.

Places.	Verfts.	Houfes and Churches †.	Date, 1786.	Time of Arrival.	Departure.
			January.		
To Kirgifhanfky Kreporft	28	60	17.	12. 45	14.
Diogroboffky do. -	23	100		17. 40	21.
Belimboieffky Savod -	23	† 1 300	18.	23.	1.
Refhotti, village -	26	† 1 10		6.	6. 15
Ekaterineburg, city -	21			11.	
			19.		8.
Kofulina, village -	24	24		12.	12. 45
Belojarfk -	24	30		17. 30	18.
Belifki, village -	25	† 1 30		21. 30	22. 30
Chornoi Korova	24	30	20.	2. 15	3. 15
Kamifhlov, city -	27	200		7. 30	8. 30
Bufhminfka, town	32	† 1 50		13. 50	14. 30
Kujarffky, village	14	† 1 15		17.	17. 30
Beloi Jalamfky -	14	40		19. 30	20. 15
Bela Kaffka, town -	28	50	21.	1. 15	2.
Tugulunfka -	32	† 1 60		6. 30	7. 30
Ufpianfk -	22	† 1 80		12. 45	15.
Tumen, city -	30	† 1		20.	
Kafkara, village -	24	20	22.		1.
Sofonov -	22	45		3.	4. 20
				9.	9. 25
Kofmakof -	21	† 1 30		14. 15	14. 30
Prokoffky -	10	150		15. 45	16.
Ufolka -	12	† 1 80		18.	18. 15
Jarkova -	15	40		19. 30	19. 45
Artamenof -	9	20		20. 30	21.
Jevleff -	10	18		22.	22. 30

Places.	Verfts.	Houfes and Churches †.	Date, 1786.	Time of Arrival.	Departure.
			January.		
To Antepena, village -	18	30	23.	1.	1. 15
Lipoffky -	17	30		2. 30	2. 45
		† 1			
Baikaloff -	18	70		4. 20	5.
Turbayeffky, tartar huts	23	40		8.	10.
Rechkoffka, village -	25	7		13.	16.
Tabolfk, city -	25			20.	
			24.		19.
Bakfheva, village -	29	10	25.	24.	1.
Stara Pogoft -	20	30		3.	4. 40
Kapotilova -	30	8		8.	9.
Drefvanka - -	31	6		16.	16. 45
Iftitfki Jurti -	30	14		21.	21. 45
Kuferadfka, village	41	30	26.	5. 30	6.
Golopopova -	56	30		17.	18.
Vikolov, town -	31	75		22. 30	23. 30
		† 1			
Otfhimova, village	46	30	27.	5. 20	6. 20
Zudiloffky, for poft -	58	30		17. 25	18.
Ribina, village -	36	20		22.	23.
Chaoonina -	40	20	28.	4.	7.
Leffka, town -	30	70		12.	12. 30
		† 1			
Butakova, village -	38	40		18.	19.
Tara, city -	29			21.	
			29.		2.
Uftara, village -	32	50		5.	6.
		† 1			
Refhetnikoff, village	36	30		11. 15	12.
Artin -	14	30		14. 30	15.
Refina - -	52	80		21. 30	23.
Marafhi -	24	20	30.	2. 30	3.
Nazareva -	12	80		4.	4. 15
		† 1			
Ghochlova - -	19	70		6. 30	7.
Voznefenfka, town	20	100		10. 30	11.
		† 1			

Places.	Verſts.	Houſes and Churches †	Date, 1786.	Time of Arrival.		Departure.
			January.			
To Tartarſki, village -	20	35		14.	30	15.
Turomova -	21	80	-	18.	30	19.
Pokrofſka, town	17	70.		21.		21. 45
		† 1				
Antofkin, village -	23	50		24.		24. 15
Bulatova -	18	75	31.	4.		4. 30
Kainſk, city -	33	125		7.		7. 20
		† 1				
Oſſinova Kolki, village	29	70		23.	45	24. 15
Kolmakov -	30	50	Feb. 1.	3.		4. 15
Ubinſky -	30	50		8.		9.
Kargan -	28	50		12.		12. 15.
Karbotſki, for poſt -	26	70		15.	45	16.
Kirgatſki Dubrovi -	25	50		24.		24. 30
Itkula, town -		100	2.			2.
		† 1				
Sektinſky, village -	47	66		5.		5. 20
Oftſhinikoff -	17	50		7.	45	8.
Sheligino -	26	70		11.		11. 15
Taraſhinſka -	20	30		14.		14. 15
Tſhauſhſka, town -	25	80		17.	15	17. 40
		† 1				
Dubrovina, village·	40	8		21.	30	22.
Ajaſhinſk -	35	40	3.	2.		2. 45
Karaſina, village -	25	10		5.	30	7. 30
Chornoi Kaſtanits	33	20		10.	30	11. 5
Varuchina -	34	60		15.	20	15. 30
Kaltai - -	23	23		17.		17. 30
Tomſk, city -	24			19.	45	
Semenuſhni, town	28		4.			23.
		35	5.	2.	30	3. 35
		† 1				
Chaldeiſki, village	14	40		4.	15	5. 25
Turuntaiva -	22	40		8.	30	9.
Cleon -	40	30		17.	30	19.
Potſhitanti -	22	20		22.		22. 20
Berikul -	25	24	6.	2.		2. 30
Kiſkova, town -	52	60		12.	30	13.
		† 1				

Places.	Verſts.	Houſes and Churches †.	Date, 1786.	Time of Arrival.		Departure.	
			February.				
To Suſlova, village	23	15	6.	14.	30	15.	
Tezſhin -	28	60		18.	30	19.	
Itat - -	32	30		23.		24.	
Bogotolſki, town -	34	250	7.	4.		4.	20
		† 1					
Kraſnoreka, village	28	150		6.	45	7.	45
Archin, city -	28	150		11.	30	12.	
		† 1					
Chornoi-rechka, village	32	60		16.	45	17.	
Bolſhoi Kemtſhuk -	38	40		22.		23.	
Maloi Kemtſhuk	35	26	8.	5.		5.	30
Zavedeva -	31	30		8.	30	9.	
Kraſnojarſk, city -	25			12.		14.	30
Botoi, village -	24	102		17.		17.	20
Kuſkun, village -	23	35		21.	45	22.	
Balai -	32	30	9.	1.		1.	40
Jarr -	24	60		5.		5.	30
Ribnia -	25	80		9.		9.	30
		† 1					
Klutch -	30	30		13.	15	14.	
Uria - -	20	30		17.	30	18.	
Kamſkoi Oſtrog -	25	70		21.		21.	30
		† 1					
Ilan, village -	20	15		24.		24.	30
Poim -	28	20	10.	5.	30	6.	
Tini - -	18	10		8.	30	9.	30
Klutchi -	28	5		14.	45	17.	30
Beruſa, town -	40	60		22.		22.	30
		† 1					
Bayronoff, village	21	20	11.	2.		2.	45
Rozgonia, hut -	24	1		6.		6.	45
Alzamai, village -	19	8		10.	30	11.	
Zamſor -	32	10		15.		15.	30
Ook - -	35	10		20.		20.	30
Udinſk, city -	25	300		22.	20	23.	15
		† 2					
Singui, village -	30	12	12.	2.		2.	15
Shabatan	30	16		6.		6.	30

Places.	Verſts.	Houſes and Churches †.	Date, 1786.	Time of Arrival.	Departure.
			February.		
To Toolon, town -	46	130	12.	11.	11. 30
		† 1			
Saragool, village -	26	30		13. 30	14.
Kuilton -	41	70		18. 30	19.
Kamelte -	38	120		22.	22. 30
Zeminſk, town -	28	50	13.	1. 15	2.
		† 1			
Dolroi -	46	80		6. 15	7.
		† 1			
Kupulin •	29	100		10.	10. 15
		† 1			
Chiremchova -	28	80		12. 30	13.
		† 1			
Tatook -	38	110		16. 15	17.
		† 1			
Viligtui -	36	70		19. 30	20. 30
Irkutſk, city -	24		14.	1.	6.
			May 10.		
Kuda, town -	18	200		8.	18.
		† 1			
Oyok -	13	180		22.	22. 10
		† 1			
Buſinſki -	27	1.	11.	24. 30	1.
Olonſki -	30	1.		4. 50	5.
Bayendarſk -	30	1.		7. 15	7. 25
Kudunſa -	30	1.		10.	15. 30
Manſurka -	30	1.		19.	19. 30
Iſiet -	30	1.		22.	23. 30
Katchuga Priſtan -	22	15.	12.	2.	

The following Villages we paſſed on the River Lena.   Verſts reckoned from Katſhuga Priſtan.

Places.	Verſts.	Date, 1786.	Houſes.	Churches.
		May.		
Katſhuga, town - -	3	15.	70	1
Kiſhnova - -	14		10	
Vircholenſk - -	30	ſtaid till 16.	100	2
Unitſki -	35		5	
Kulioſſki -	54		6	
Kozloſſki - -	59		5	
Apuſhinſk - -	64	waited day-	6	
Kaſhinoſſki -	80	light.	16	
Zapleſhinſk - -	86	-	8	
Garaſovoi - -	90		3	
Pagoſſki - -	97		21	
Verobroſſki - -	100		14	
Mikiſhinſk - -	107		6	
Apaſhinſk - -	119	17.	5	
Golovna - -	123		6	
Ardoſſki -	126		11	
Golovnoſſki - -	128		10	
Kuznetſoſſki - -	131		6	
Simeonoſſki - -	132		7	
Balaganſkoi - -	135		12	
Ziranoſſki - -	137		7	
Gigaloſſki - -	140		13	
Uſtilga - -	170		50	1
Gruſna - -	194		9	
Botoſſki - -	218		10	
Shamanoſſki - -	228		15	
Golli - -	244		8	
Sherſtinova - -	251	18.	1	
Starſa - -	254		2	
Tomſkina - -	257		7	
Saroſſki - -	264		3	
Zagobininſki - -	284		9	
Baſoſſki - ..	299		7	
Dudkin -	303		9	
Orlinga - -	324		20	1

Places.	Verſts.	Date, 1786.	Houſes.	Churches and Monaſteries †.
Povoſki - -	325	May 18.	15	
Poolioffki - -	327		11	
Viſoka - -	334		7	
Taraſova -	342		8	
Sedunoffki -	344		8	
Scoknioffki -	364		4	
Boyarſki -	384		8	
Pavlova -	394		1	
Omoloffki -	407		6	
Sinuſhkin - -	413		3	
Riga -	420	19.	3	
Turoka - -	454		8	
Uſkoot - -	469		20	1
Balachaia -	480		9	
Yekurin - -	487		10	
Polovinoi -	498		3	
Podimachinſkai -	512		14	
Koſarki -	513		10	
Kokooiſhka -	531		4	
Ti-oora -	541		10	
Nazaroffki -	584		10	
Marakoffka -	601		20	1 † 1
Tyra - -	615		10	
Ulgan -	623		8	
Kaſemeroffki -	633		3	
Kraſnoyarof -	641	20.	4	
Levonoffki -	649		3	
Potapoffki -	655		6	
Luboffki -	674		4	
Karaſoffki -	678		4	
Sheſtakova -	683		2	
Gaviloffki - -	684		2	
Panſkoi -	685		5	
Balaſhova -	688		2	
Makarova -	690		11	1
Panſhina -	702		2	
Zaborſkoi -	705		12	
Krivalutſkoi -	710		20	1
Vologinſk -	713		5	
Lavruſhinſk -	718		4	
Lazarova -	725		5	
Menakoffkoi - -	730		2	

Places.	Verſts.	Date, 1786.	Houſes.	Churches and Monaſteries †.	
		May 20.			
Tſhertoſſkoi - -	745		2		
Kudrina - -	750		3		
Kulibakinſk - -	766		4		
Krivorotſkoi - -	772		15		
Kabarova - -	776	21.	10		
Varoninſk - -	777		15		
Kiringa - -	778		80	3	† 1
Nikolſki - -	780		4		
Zme-ina - -	790		9		
Alexeeſſki -	802		8		
Soltikoſſki -	815		19		
Podkaminoi -	818		20		
Polovinoi - -	821		12		
Pikulina - -	824		4		
Borovia - -	825		2		
Meſovia - -	826		3		
Gerbova -	828		20		
Banſhikoff - -	830		20		
Tſhigioſſki - -	838		25		
Grebenie -	840		10		
Kondraſh inſk - -	844		12		
Veſnikova - -	854		20	1	
Sukniova -	869		20		
Spoloſhna - -	874		30	1	
Kabalova - -	875		20		
Puſhinova - -	894		10		
Ilienſk - -	899		7		
Darinſk - -	919		10		
Izſhura - - -	949		11		
Davidoff -	963	22.	7		
Korſhunoſſki - -	980		7		
Ivanuſhka - -	981		3		
Tſhaſtinſk -	1004		2		
Varobieſſki - -	1041		1		
Kuraiſk - -	1070		4		
Poiſhina - - -	1110		3		
Tſhuiſki - -	1157		7		
Vitima - -	1178		30	1	
Pelidui - -	1205	23.	30	1	
Kriſtoſſki -	1232		3		
Yeloſſki - -	1259		4		
Pirkinſk -	1293		4		

Places.	Verſts.	Date, 1786.	Houſes.	Churches and Monaſteries †.
Ghamra - -	1309	May 23.		
Konki - -	1336		3	
Tſhiooſka - -	1380	24.	2	
Muria - -	1405		3	
Silguil - -	1445		3	
Newye - -	1475		5	
Yerba - -	1505		5	
Ooſhakan - -	1540		3	
Yedai - -	1565		2	
Maekai - -	1595	25.	4	
Beroſova - -	1645		2	
Dolgoi - -	1678		4	
Nelena - -	1711		3	
Cheringa - -	1736		3	
Birt - -	1775		4	
Anyinſk - -	1798		6	
Alofinſk - -	1799		9	
Olekma - -	1807	26.	13	
Solenka - -	1832		50	
Namania - -	1872		4	
Karabalyk - -	1912		3	
Chatin Tumul - -	1954	27.	3	
Murta - -	1976		2	
Sanayagtak - -	2018		1	
Malikan - -	2055		2	
Iſaki - -	2090		2	
Nevarchie - -	2125		2	
Umarie - -	2150		2	
Sinae - -	2180		2	
Batamai - -	2210		2	
Kitarie - -	2232		3	
Toiona - -	2274		2	
Biſtach - -	2301		2	
Yakutſk - -	2390		14	

No. IV.

====

ACCOUNT

OF

The full Pay of the different Ranks, with other Dependencies; and also an Explanation of the usual Deductions, according to the Regulation of 1782.

	Ro.	Co.	Ro.	Co.	
Captain of 1st rank	600				
Allowed 6 Denshicks; their pay each - -	6				
Which pay is understood for all Denshicks of others					
Captain of 2d rank -	420				
4 Denshicks					
Captain-Lieutenants, Majors' rank - -	300				
3 Denshicks					
Lieutenants, Captains' rank -	200				
2 Denshicks					
Midshipmen - -	120				
1 Denshick					
Upper Auditors -	240				
2 Denshicks					
Auditors - -	100				
1 Denshick					
Clerk, or Secretary -	72				
Skippers					
Of the 1st rank -	144				
1 Denshick					
Of the 2d rank -	132				
1 Denshick					
Commissary -	100				
1 Denshick					
Priests -	120				

	Ro.	Co.	Ro.	Co.	
Surgeons					* According to their merits and abilities, their pay to be augmented or diminished.
Each allowed 1 Denshick {	300				
	240				
	180				
Ships' Clerks -	150				† According to their attention, abilities, and desert, to add or deduct; but never less than ro. 60, nor more than their full pay.
Sturmen - - *	36				
Pod Sturmen	138	* 40			
Pod Lekars -	60				
Pod Skippers -	84				
Timmerman (Ship Builders) †	60				‡ According to merit, their pay to be increased or lessened; but never to be less than ro. 24, nor to exceed their full pay in addition.
Boatswains -	90				
Boatswains' Mates -	60				
Sturmens' Learners - ‡	36				
Surgeons' Learners	31	50			
Quarter-Masters	18				
Sailors { 1st - - -	24				
Sailors { 2d - -	11	14	5	36	} Their uniforms to be given in natura.
Cabin Boy - -	7	64	5	36	
Desatnick of Plotnicks -	6	9	4	16	
Plotniken -	24				
Caulkers -	15				From 12 ro. to 18 ro. according to their merit; but not to exceed this stipulated sum.
Sail-Makers' Mates -	15				
Smiths -	15				
Coopers and under Coopers	15				
Boteleirs -	15				
Under Boteleirs -	24				
Trumpeter, 1st and 2d Clafs §	11	14	5	36	Their uniforms to be given in natura.
Kettle Drummers	40				
Cooks 1st and 2d Clafs	60				§ According to their knowledge of music and good behaviour their salary may be augmented or diminished.
Profort - ‖ {	9	14	5	36	
In the Sea Hospital	6	14	5	36	
Doctors -	800				Their uniforms in natura.
3 Denshicks					
Stab Lekars -	600				
2 Denshicks					
The Upper Priest of the Fleet above Church Characters	240				
Marines					
Major Præmier	300				

N. B. Denshick is a Servant allowed out of the Ship's Company, not only whilst at sea, but also at quarters. This man may be let out to work, and the Officer receive the money that he gets by labour, as also his allowance of provision.

INSTRUCTIONS

OF

HER IMPERIAL MAJESTY,

FROM THE

ADMIRALTY COLLEGE,

To Mr. JOSEPH BILLINGS, Captain-Lieutenant of the Fleet, commanding the Geographical and Aftronomical Expedition intended for the North-Eaftern part of the Ruffian Empire.

HER Imperial Majefty, extending her maternal and unremitted care for the happinefs of her fubjects to all, even the moft diftant, parts of her vaft dominions, has been gracioufly pleafed to order, as well with intent to furnifh them with better means of life, and to render them more happy and advantageous, as for the important advancement of fcience, an expedition of difcovery to the moft eaftern coafts and feas of Her Empire; for the exact determination of the longitude and latitude of the mouth of the river Kovima, and the fituation of the great promontory of the Tfhutfki, as far as the Eaft Cape; for forming an exact chart of the iflands in the Eaftern Ocean extending to the coaft of America; in fhort, for bringing to perfection the knowledge acquired under her glorious reign, of the feas lying between the continent of Siberia and the oppofite coaft of America.

The execution of this Her Majefty's intention is entrufted to you, as a fkilful officer zealous for the fervice of Her Imperial Majefty; in full confidence, that the importance of this bufinefs with refpect to the glory of Her Majefty's facred name, and the intereft of Her Empire, will excite you to fulfil the great expectations entertained of your abilities.

Her

Her Imperial Majefty, agreeably to her wonted gracious and generous difpofition in all her ufeful and maternal commands, is pleafed, over and above fuch
weighty incitements, for your greater encouragement to activity and zeal in the
fervice, to give you the rank of Captain-Lieutenant of the fleet; for which rank
you have taken the oath, and received your patent; and, to favour you ftill more,
the officers and petty officers which you have demanded are named according to
your own choice, as you will obferve by the lift annexed hereto.

At the fame time Her Imperial Majefty has gracioufly ordered, that from the
day of figning this Inftruction, until your return to St. Peterfburg, you and all
under your command are to be allowed double pay, according to their ranks; to
you according to the rank here granted, and to your fubalterns according to the
rank that they fhall obtain at Irkutfk; which pay is to be given here, one year in
advance; above which, to you and all your fubalterns, a bounty of one year's pay
for procuring neceffaries for travelling.

Our Moft Gracious Sovereign has alfo generoufly ordered, that at your arrival at
Irkutfk, before you begin the execution of what is prefcribed in the following articles, you fhall declare in Her Majefty s name, to all officers and petty officers
under your command, an advanced rank above what they bear, and have them
fworn accordingly; except thofe only who are to receive gratifications in money,
according to the annexed lift.

Her Imperial Majefty gracioufly orders you to declare yourfelf, in Her Imperial
name, Captain of the Fleet of the fecond rank, after having fulfilled the bufinefs
prefcribed in the following articles on the river Kovima; in which rank you are
then to take the oath.

When you have finifhed your prefcribed bufinefs on the river Kovima, and along
the coaft of the Tfhutfki, at your return to Ochotfk, where every thing will be
ready for your voyage to the coaft of America, at the inftant of going on board
you are to declare, in Her Imperial Majefty's name, an advanced rank to all under
your command; to caufe the oath to be adminiftered to yourfelf, and to the reft
according to the above-mentioned lift. Laftly, at your arrival at Cape St. Elias
you may declare yourfelf Captain of the firft rank.

Thofe of your fubalterns who, according to their rank, fucceed to the places of
fuch as may die, either a natural death or by accident, and who will be ordered to
fuch rank either by you or by the officer that may have the command after you,
provided they produce a certi cate of their good behaviour and zeal in the fervice
from the Commander in Chief, will on their return to Peterfburg be confirmed at
the Admiralty College, in the name of Her Imperial Majefty, in the rank conferred on them; and will be accounted in that rank from the day of their appoint-

3

ment.

ment. This is to be underſtood of thoſe who bear petty officers' ranks; thoſe who get into the denomination of upper officers, according to the above-mentioned order of advancement, will have equal advance with officers that go from hence.

In caſe any one of thoſe that go from hence ſhould die, be maimed, or loſe the uſe of his limbs, during the Expedition on the Tſhutſki coaſt, or the navigation from Ochotſk to the American coaſt; if ſuch perſon ſhould have a wife and children, the widows of the deceaſed ſhall receive until they marry again, or until their death, and the children till they come to their lawful term of years, half pay of what the deceaſed received during the Expedition; the maimed ſhall alſo receive ſuch half pay during their lives.

After having completed the buſineſs entruſted to you, on your happy return to St. Peterſburg, you, and all under your Command, will receive the defect of the double pay for the different ranks obtained during the Expedition; and, as a gratuity, a year's double pay according to the rank they return in; over and above which you and all your ſubalterns, returning ſafe, will receive for life the ſingle pay received during the Expedition, without accounting for what he may get for future ſervices.

Such gracious grants and further promiſes of protection, but moſt of all the importance of the truſt laid upon you, muſt excite in you a noble emulation to render yourſelf worthy of it, by endeavouring to do all in your power to fulfil the articles of this inſtruction, confirmed by Her Imperial Majeſty, and ſetting, by your unremitted endeavours, an example of zeal to all your ſubalterns.

ARTICLE I.

For your information are hereunto annexed fourteen charts of former navigators on the Northern and Eaſtern Ocean, and along the coaſts; as alſo of travels by land; to which are annexed ſhort extracts of the journals of the travellers, from 1724 to 1799. The plan of the veſſel preſented by you for inſpection is herewith returned; and you may, upon that plan, conſtruct veſſels at Ochotſk, if there be not one found there fit for your navigation. Annexed is likewiſe a liſt of Ruſſian towns, with the determination of the latitude and longitude of ſome; as alſo a model, according to which vocabularies of the different nations are to be collected. You receive alſo medals expreſsly made for you, to be employed with ſuch nations, the proper appropriation of which will be hereafter deſcribed.

You will receive herewith five thouſand rubles, to be employed in buying beads, knives, and other inſtruments, ſmall copper-kettles, and other ſuch trifles, to be employed as preſents to the ſavages who are fond of them.

You

You will also receive here mathematical and astronomical instruments, besides others; and double pay, for you and all your Command, for one year advance; and likewise the above-mentioned bounty granted by Her Imperial Majesty for you and all your Command, which you are to deliver against their receipts in the official receipt-book, of which twenty are given to you from the Admiralty for this purpose; as also for entering for the future all receipts and expences. After you have provided yourself with all necessaries for the journey, you are to proceed with all your Command, the shortest and most advantageous road to Irkutsk. You are to take care not to break your oath of keeping secret the business entrusted to you; and not to exceed, on affairs of secrecy, the ukaze of 1724, of which a copy is annexed for your information. You are not to open yourself on any account to any body about the measures or proceedings of your Expedition, unless ordered so to do; and much less so, to any body, this or any other instruction that may be given to you for the same purpose; you are also to give the most strict orders to all your Command to this effect.

During your travels, if any very important accident should happen to you, you are to give notice to the Admiralty College by express; but in affairs of less importance, for example of the state and place in which you are, send your reports by post. From the day of your setting out from Petersburg till the very conclusion of your Expedition, you are to keep a journal very accurately yourself, and order your officers to do the same.

ARTICLE II.

When you arrive with your Command at Irkutsk, you are to deliver to the Governor-General of Irkutsk and Kolivan, Jacobi, or in his absence the Vice Governor, the original ukaze of Her Majesty directed to him; to which is added a copy of this your Instruction; and in which order is given, that all possible assistance be rendered at your request for the service of Her Majesty. The Governor is to give you sufficient directions for your journey to Yakutsk, Ochotsk, Izshiginsk, and to the river Kovima. He is to provide you with an open ukaze, by which it is enjoined to all the commanders and chanceries of the places through or by which you, or any sent by you (to whom you are always to give at their setting off your instructions for their journey), may travel, that they, upon your request, give you all possible assistance, as well of hands as stores and provisions; besides, the same Governor-General is empowered by Her Majesty's ukaze to give you another open ukaze, for the receipt of ten thousand rubles for unexpected and extraordinary expences, which may happen during your travels; as also for travelling expences, and for the payment of such men as you may, according to the prescription of this instruction, employ in any part of Irkutsk. Of this sum you may receive as much as is neces-

sary,

fary, and when and where you think fit; but for the money received, you, and the eldeft officer next you, are to pafs your receipts, that you may know how much money is received and can be received on the ukaze. You are to require in each place where you take money, that he from whom you receive it fhould endorfe upon the ukaze, how much, where, and when, the payment has been made; and the expences, with an account for what the expenditures have been made, are to be noted in the official book given by the Admiralty, with receipts wherever they can be procured. Stores and provifions you are to receive, with confent of your fubalterns, mentioning, in the receipt which you give, the quality and quantity of goods received. You are not on any account to make any fuperfluous or puzzling demands, only what is prefcribed, or fuch as contribute in reality to the fervice of Her Majefty; nor expend any fum upon what is not neceffary, as you will be refponfible for it.

At Irkutfk you are to endeavour, with the help of the Governor-General, to provide yourfelf, without the leaft lofs of time, with all neceffaries, and to get them tranfported to their refpective places. You may, for forwarding bufinefs, detach from your Command upper and under officers for infpecting, preparing, and tranfporting the ftores collected to their places of deftination.

If you fee, by the lift in the poffeffion of the Governor-General, that in the magazines at Ochotfk there is not a fufficient quantity of provifions and other ftores neceffary for duly arming and victualling the fhips which are proper to be employed for your navigation, as alfo for your march to the river Kovima, and along the coaft of the Tfhutfki; in fuch cafe you are to requeft the Governor-General to endeavour by all means to furnifh the magazines in due time with what is requifite, and that the faid Governor-General may fend an exprefs to the Commander of Ochotfk, with orders as well to fupply fuch wants, if there fhould be any, as alfo to ftop the veffel that annually fails with provifions for Izfhiginfki Krepoft in June or July, that you may be able to take the opportunity of faid veffel for going to Izfhiginfk; and laftly, that the faid Commander fhould fend orders to Petro Pavloffky, or whatever other harbour of Kamtfhatka is thought more proper, for preparing there, againft your intended voyage towards the coaft of America, fufficient quantities of dried fifh and wild roots, and other eatable wild vegetables, for the fupply of your people; enjoining, that at fuch harbour fhould be ftationed in due time about twenty Kamtfhadals, ufed to a feafaring life, and well fkilled in fifhing and hunting, who are to accompany you in your voyage for the ufual pay.

At Irkutfk, you may examine and take your choice of five or fix of the beft fcholars of the Navigation School, and take them under your command to employ them during your travels in furveying and drawing charts: thefe are to remain

with

with you till the conclusion of the Expedition, upon the same footing as the other petty officers that go with you from Petersburg. Those Uchenicks that were formerly sent with Captain Krenitzen received fifty-four rubles annually; you may give them such payment for one year for their equipment.

You are also to take with you from Irkutsk the naturalist Mr. Patrin, who will remain with you till your return with your Command to St. Petersburg, in order to describe such natural curiosities as may be met with during the course of the Expedition: he will receive particular instructions for his business, and what he is to do in such places where he will go with you, or where you shall think fit to send him, for describing objects worth observing; you are to assist him, upon his request, with hands, instruments, and money for executing his orders; giving him leave to stop for observations in such places so long as circumstances will permit, taking him along with you wherever you go to distant places. You may, if you shall think it necessary, receive from the Governor-General at Irkutsk, according to the imperial ukaze, another year's double pay for all your Command in advance.

Having received from the Governor-General all that is required for the Expedition, and all that may serve for your future and more circumstantial information; having also executed all that is to be done at Irkutsk, and reflected on circumstances that may happen during your further journey, you will then, without loss of time, either by land or along the river Lena, as you shall think best, with such of your Command as remains with you after making the necessary detachments, proceed to Yakutsk, or where you shall think it most convenient for the service, or the intent of the Expedition. As you are strictly to follow the directions of the Governor-General, so you have also to make your reports to him of your proceedings, of unforeseen untoward circumstances and hindrances in your journey to Ochotsk, and from thence to Izshiginsk and to the Kovima; in order that you may, in case of necessity, receive directions from him how to proceed.

Lastly, You are to represent to the Governor-General, that he is to give the most absolute orders through his whole government, that nobody should be curious in opening letters sent by messengers with private reports, as it happened during the Expedition under the command of Captain Krenitzin the 10th of April, in the year 1768, at the port of Ochotsk, by the Commander Colonel Feodor Plenisner.

Particularly at this time, and in this part of the Russian Empire, most of all in parts lying beyond the river Lena, as far as you shall travel either by sea or land, you are to determine as nearly as possible the longitude and latitude of remarkable places, the variation of the compass; to form surveys and charts; draw remarkable views of coasts, with the situation of bays, inlets, and roads; and mark their advantages for trade, fisheries, &c.; likewise to observe and describe the time, strength,

rising,

rifing, and irregularity of tides and currents; alfo of rocks under water, fhoals, and other dangerous places; the ruling, variable, and trade winds; the changes of weather; meteors, particularly Aurora Borealis; the ftate of the electricity of the air during thefe meteors, and their influence on the compafs; laftly, the changes of the barometer and thermometer.

Moreover, Mr. Patrin will have particular inftructions refpecting his obfervations in natural hiftory: however, you are never to neglect, efpecially when he is not prefent, to obferve the nature of the foil accurately, and of the productions of the country where you find yourfelf; you are diligently to collect feeds, ripe fruits, and dried plants, branches and pieces of the wood of remarkable trees, their barks, refins, and gums; alfo fea-weeds, zoophytes, fhells, fifhes, amphibious creatures; infects, birds, and other animals; taking off and ftuffing the fkins of fome, and drying and preferving in fpirits others. You are likewife to collect fpecimens of ores, foffils, ftones, falts, earths, and fulphurs; noting the place where each were found or caught, and at what time.

To prevent fuch collections being fpoiled by accidents, you may leave them in fuch places as you think proper, where you may take them up at your return to St. Peterfburg. If in fuch places there fhould be a commander, you are to deliver them to him, taking a receipt. If the places are not inhabited, put them in remarkable fituations, where they will be fecured from weather and deftruction; or, ftill better, fend them along with your reports and their defcription, under your feal, to the Governor-General of Irkutfk.

You are likewife to make, if poffible, circumftantial defcriptions of the quality and ufe, and even drawings of the moft curious productions of nature; you are to enquire accurately about the number, ftrength, natural difpofitions, manners, and occupations of the inhabitants of unknown places; likewife order to be made vocabularies of their language, after the model given you; endeavouring to exprefs as nearly as poffible the pronunciation of their words in Latin and Ruffian characters. Laftly, you are to procure, (or, if that be not poffible, to get painted, or defcribe) the furs, dreffes, arms, and manufactures, of fuch nations.

ARTICLE III.

Upon your arrival at Yakutfk, you are to apply yourfelf immediately, to execute what the Governor-General may think neceffary to prepare for your further journey to Ochotfk; and during your ftay there, by virtue of your open ukaze, which orders all Commanders and Gorodnitfhi of the towns through which you pafs, to give you all neceffary affiftance, you are to require abftracts of accounts to be

　　　　　found

found in the archives of late navigators, and of all that can give information about your main buſineſs upon the Kovima, and round the coaſt of the Tſhutſki ; and if you find by ſuch liſts or abſtracts that there is any thing ſurpaſſing the extracts communicated to you at St. Peterſburg, and you think them neceſſary, you may demand copies of them ; and if there be any charts get them alſo copied.

Wherever you produce the open ukaze of the Governor-General of Irkutſk, you may permit to ſuch perſons as it regards to take copies of it, in caſe it ſhould be neceſſary.   You may, if you and the Governor-General ſhould think it convenient to be done at Yakutſk, and not by preference at Ochotſk, Izſhiginſk, or even the Oſtrogs upon the river Kovima, pick out the neceſſary number of Coſſacks, ſoldiers, interpreters, and guides, chooſing preferably hunters, and ſuch as are recommended for their ſkill and good behaviour, and who have been upon the Kovima ; and of ſoldiers ſuch as were formerly in garriſon at Anadirſk, have converſed with the Tſhutſki, frequented their habitations and the environs of the Kovima, and the coaſt of the Frozen Ocean (ſome even were born among the Tſhutſki) ; with theſe people you may, in preſence of the Commander of the town, either make an agreement, or pay them without agreement, double the ſum that is uſual there for people who are hired for a term to ſerve at ſea ; which they are to receive from the time you take them under your command, till you diſmiſs them at the cloſe of the Expedition, or till their death, inſcribing this pay in a particular official book ; and you may promiſe in the name of Her Majeſty, to ſuch as offer themſelves volunteers, that at the happy return from the Expedition they ſhall receive a gratuity of one year's pay, as received during the Expedition, for their ſervice.

Following the example of your predeceſſor Captain Krenitzin, who was ſent in 1764 to theſe ſeas, you may, if you think it conducive to the ſervice, and for more expedition, which in all your proceedings is hereby much recommended to you, order at Yakutſk (as he did in 1765) rope work to be tarred, and proviſions packed in bags and caſes, each containing no more than two poods and a half weight ; and when you have got the neceſſary quantity of proviſions in readineſs, ſend part of them off, under command of an officer inſtructed by you, and furniſhed with all neceſſaries, loading on each horſe no more than five poods, on account of the many bogs, rivers, and mountains, which are to be paſſed.   Yourſelf may follow in the ſame manner with the reſt of the proviſions, ſtores, and men.   To prevent hindrances on the road to Ochotſk, you may deſire the Commandant of Yakutſk to ſend off an expreſs, preparing neceſſaries for your journey.

ARTICLE

ARTICLE IV.

When you have furnished yourself with all neceſſaries at Yakutſk, you muſt make your diſpoſitions to complete the tranſport which is already ordered before, of proviſions neceſſary to maintain your party during your ſtay upon the Kovima, and the coaſts of the Frozen Ocean. If you ſhall think it neceſſary to have ſome Coſſacks to form this party, and if you can find ſuch as have been before upon this river, or upon theſe coaſts, you may make choice of ſuch either at Ochotſk, or at Inzſhiginſki Krepoſt.

ARTICLE V.

Laſtly, in order that you may beſt employ your time, endeavour to arrive at Ochotſk at the ſame time nearly with your ſubalterns, to chooſe there the ſailors and Coſſacks who are to follow your Expedition by land and by ſea. You muſt alſo chooſe, from among the pilots of that port, two or three who have ſufficient knowledge of thoſe ſeas, and whoſe ſervice you ſhall think moſt conducive to the ſucceſs of your navigation. On your recommendation, they will enjoy the ſame advantages as the reſt of your Command. Each of them is to ſelect for his aſſiſtant one of the Utſhenicks of the Navigation School of Ochotſk.

At Ochotſk you are to make all neceſſary preparations for the ſea voyage preſcribed hereafter in the 10th Article. In caſe not one of the veſſels in actual ſervice there ſhould be ſafe enough for ſuch a diſtant navigation, you muſt then take your meaſures for conſtructing two veſſels of ſufficient ſtrength and convenience, to anſwer the purpoſe and preſerve the healths of the crew. Of one of theſe veſſels, at the time of navigation on the coaſt of America, you will have the command, and the command of the other will be given to the ſecond in rank; for the ſafety of the crew, and the ſucceſs of the navigation, depend on the veſſels of the Commander in Chief being accompanied by another. In order to enforce their conſtruction, orders will be given immediately to the Governor of Irkutſk, that the beſt ſhip timber to be found about Ochotſk ſhould be prepared, and all ſtores got in readineſs for fitting out one ſhip of eighty feet in keel, and another of ſmaller dimenſions, by virtue of the open ukaze which the Governor-General of Irkutſk and Kolivan is to give. You are to demand from the Commandant of Ochotſk the neceſſary number of carpenters, and all requiſite aſſiſtance towards conſtructing and fitting up your veſſels. You are empowered to give the ſuperintendance of the docks to one or more of your ſubalterns, and to your ſhipbuilder, in order that the building may be carried on with all poſſible ſpeed,

and

and entirely according to your plan. You muft likewife order at Ochotfk a certain number of pofts of durable wood to be prepared, which are to be erected on
fuch lands as may be newly difcovered by you; thefe pofts you will ftow in your
fhip when you fail for America.

ARTICLE VI.

When you have made thefe preparations, and collected from the Archives at
Ochotfk what information and journals relative to your Expedition may be found
there, you may then without lofs of time, with part of your Command, which you
have chofen at Ochotfk, and with Affeffor Patrin, proceed on the readieft way to
the Kovima. It will be proper to go as lightly equipt as poffible on board the
veffel which fails in June or July with provifions for the garrifon of Izfhiginfk;
at that place you will find the beft Coffacks and foldiers for forming your party, as
fome of them heretofore compofed the garrifon of Anadirfk, and have had con-
nexions with the Tfhutfki, and others were even born and travelled amongft them.
With thefe you may march over to the river Omolon, down which you may float
on rafts to the Kovima. Arrived at the Kovima, you are to make geographical
and aftronomical obfervations of the latitude and longitude of Virchnoi and Neizfhnoi Kovimfki Oftrog, and the mouth of the river; and to take an accurate furvey
of it, obferving the foil and inhabitants of the adjacent country.

ARTICLE VII.

Having determined with all poffible accuracy the fituation of the Kovima, and
defcribed its courfe and the foil over which it flows, you are to endeavour, if circumftances permit, to make ufe of boats called Shitiki, conftructed as ftrongly as
poffible, to coaft along the promontory of Tfhutfki from the mouth of the Kovima
to the Eaft Cape. In cafe, however, the coafting by fea fhould be found abfolutely
impracticable, and the information received on the fpot give you hopes of reaching
it by land, you may then proceed thus to defcribe thefe coafts, going in winter over
the ice. It may happen, that by thefe means you will difcover iflands or lands
that may lie to the north of thefe coafts, and of Bering's Straits. You may continue your travels and enquiries, employing different means as far as circumftances,
fafety, and the good of the fervice, require. You are to make an accurate chart;
lay down the remarkable places that appear; take views of the coaft and remarkable
objects; endeavour alfo to get as much information as poffible of the country of
the Tfhutfki, their ftrength and manners; and, wherever opportunity offers, to
contribute by your behaviour to the fubjection of this nation to Ruffia, and to the
good opinion of the mild government to which they fubmit.

ARTICLE

ARTICLE VIII.

Whatever fuccefs the trials on the Kovima, and from thence along the coafts of the Frozen Sea, may have, after having done all that it is poffible to expect from your zeal, return from thence by the beft route to Ochotfk, to finifh there the laft preparations for your navigation in the Eaftern Ocean, to take command of the people, and of the fhips built or chofen for the Expedition. To the officer who will have the command of the fecond fhip you are to give the full complement of fturmen and failors, inftruments, ammunition, provifion, and other neceffaries for the fervice. This officer is to follow exactly your orders, fignals, and inftructions.

ARTICLE IX.

If on any unforfeen account the fhips fhould not be in readinefs, then you may, awaiting their being built, employ the fpare time and your talents in ufeful difcoveries, on the fea between the Kuril iflands, Japan, and the continent of China, even the Corea; and endeavour to bring to perfection the charts of thefe almoft unknown parts of the feas; for this purpofe, you may employ any one of the packet-boats or galliots belonging to the government at Ochotfk, which you fhall think fitteft for the fervice, and part of your detachment. This fecondary point, however, muft not make you lofe fight of the principal object of the Expedition, which you muft endeavour exactly to fulfil.

ARTICLE X.

When your fhips are perfectly loaded, armed, and provifioned at Ochotfk, you are, in company with your fecond veffel (alfo taking under your convoy the merchants' fhips that choofe), to fail in the moft favourable time for doubling the extremity of Kamtfhatka; you are to call at the port of Petro Pavloffky, or at Kamtfhatka, at whichever of the two the provifions mentioned in the 2d Article are collected. Thefe, as alfo the Kamtfhadals ordered there for the purpofe, you will diftribute to both fhips; you are then to continue your voyage for furveying the whole chain of iflands extending to America, or for the difcovery of new ones.

You are to make it a principal point of your duty to draw up an accurate chart of thefe iflands, determining their fituation by frequent obfervations; and, endeavouring to get a knowledge of the beft harbours, roads, &c. to be found on
them,

them, you will extend these enquiries even to the coast of America; and chiefly
direct your attention to the islands hitherto little frequented, and not well known,
which lie along and south of the coast to the eastward of the island of Oonimak
and the great promontory of Alaksa, which is part of the continent. Such islands,
for example, as Sanajak, Kadiak, and Lesnoi, the islands of Shumagin and Too-
manoi, seen by Bering and others.

During your navigation in these seas, if you should meet with other ships, under
English, French, or other European colours, you are to behave in a friendly manner, and not give occasion for dispute.

A R T I C L E XI.

Having usefully employed the summer in these enquiries, you may, at the setting-
in of the stormy season in autumn, look out for a proper harbour, either in America, or on the islands lying in these seas, or in Kamtshatka, there to winter and
refresh your men; and you may again continue your endeavours and enquiries
when the favourable season returns.

A R T I C L E XII.

As some indications observed by Captain Bering on his sailing towards America,
and which were confirmed by the English Captains Clerke and Gore at their return
from the Sandwich islands to Kamtshatka, give reason to conjecture that there are
islands situated to the southward of the known chain of islands, and to the eastward of the meridian of Kamtshatka, between forty and fifty degrees of latitude,
you may try, on your going, or in your return, to discover these unknown islands,
and get information respecting them, for the good of the trade of Kamtshatka; not,
however, losing too much time on these uncertain trials.

A R T I C L E XIII.

You are authorised to make enquiries about such parts of the continent of America as former navigators could not well survey on account of bad weather; chiefly
endeavouring to discover their best harbours, which may serve in time for opening
a fur trade with the inhabitants of the Continent; and in all cases principally endeavour to get a knowledge of the different productions of the Continent, islands,
and adjacent sea, as prescribed by the 2d Article.

ARTICLE XIV.

For this reason you are to give the naturalist, Mr. Patrin, whenever he requires it, full liberty, with necessary assistance, and furnish him with opportunities to do his duty; leaving him on shore as long as the service permits in such places as he may think worthy of observation, or sending to such places as he shall indicate. The observations, enquiries, and collections, which he will make in his way, you are to leave at his entire disposal till your arrival at Petersburg, whither he also is to return to deliver them.

With respect to every thing that regards the reports which you are from time to time to make during your Expedition, he may also avail himself of the same opportunities, and deliver you extracts and results of his observations, in the language wherein he may be able most clearly to express himself. If by any accident, or illness, Mr. Patrin should not be able to prosecute his researches, then you are to take care of his manuscripts and collections, sealing them up in the best condition till he recovers, or, if necessary, till your return.

ARTICLE XV.

On such coasts and islands as you shall first discover, whether inhabited or not, that cannot be disputed, and are not yet subject to any European power, you are, with consent of the inhabitants (if any), to take possession in the name of Her Imperial Majesty the Sovereign of all the Russias, of the places, harbours, and all advantages which you think useful, in the manner prescribed in the following Article.

ARTICLE XVI.

When you bring under Russian subjection newly-discovered and independent nations, or people, you are to observe the following directions. As such people have most probably never been insulted by any Europeans, your first care must be, chiefly to give them a good opinion of the Russians. On finding such a coast, island, or promontory, you are to send one or two baidars, with armed men, under command of an experienced sturman, with interpreters, and small presents with them. Let them look out for a harbour or bay to secure your vessels in; when such are found, take the soundings and go in; but if such harbour cannot be found, you may then send baidars, or boats, with part of your Command, on shore,

shore, to examine if there be inhabitants, forests, animals, &c. They are not to land all together, but leave a guard over the boats; and the landed party are not to scatter, but keep together. If there are inhabitants, they are to speak to them by interpreters, who are never to be sent alone, but accompanied by some men secretly or openly armed; for it has happened, that savages have killed or carried off interpreters, to the no small disappointment of the discoverers. The interpreter is to speak to them, as from himself, of your friendly intentions; to shew which, he is to give them choice of presents, entreat them in a friendly manner to accept of them, and invite the chiefs on board the ships; to flatter them, give them medals to hang about their necks (which are delivered to you for the purpose); tell them, that these medals are a token of the lasting friendship of the Russians; ask and take from them what they choose to give as the like token; persuade them to tell all their countrymen that the Russians wish to be their friends; enquire their name, and the origin or meaning of it; whether their population is numerous, particularly in males; ask concerning their religion; their idols (respecting which you must carefully observe that none of your Command go near or destroy them); their food and industry; where they travel, and by what means; how they call the places to which they resort, and on what point of the compass they lie, whether islands or continent; and when they point out the situation with their hands, observe secretly, but accurately, the situation of the compass, and note in the journal how far distant; if you do not understand their measurement, ask how many days' journey or voyage, that you may know how to keep your course, if you think it necessary to go there; also ask if there are on such coasts or islands any considerable bays; whether large ships with one, two, or three masts and sails frequent them, or whether such ships do not frequent their own or neighbouring islands, or coasts? If you see in their hands any article of European or Asiatic workmanship, ask whence they had it; make all necessary observations for the description of the place, and ask their permission to come often on shore; learn their custom of saluting each other, and salute them so when you meet. When they come to like you for your friendship and generosity, and you are sure that they are not subject to any European power, then tell them that you have a mind to look out for such other friends; and that they may permit you, as your friends in other places do, to erect a mark on some high place on shore, by which you may again find out the place where the friends of Russians live, and that this should be done, according to your custom, with ceremonies; when they give this permission, then order, upon one of the posts prepared at Ochotsk, marked with the arms of Russia, to be cut out letters indicating the time of discovery, a short account of the people, their voluntary submission to the Russian sovereignty, and that this was done by your endeavours under the glorious reign of the Great Catharine the Second.

You

You are empowered to name the iflands and countries that you difcover as you pleafe, if they have no proper name. When the poft is prepared, let the inhabitants know that you will come on fhore to fix your mark, which you are to do with proper ceremony and precaution; after which make the inhabitants prefents of fmall things which they like; and to the chiefs give medals, which they can hang about their necks; laftly, perfuade the inhabitants, that if they choofe to remain friends to the Ruffians, they fhould never permit either their own people or foreigners to dig out or fpoil this mark, but preferve it entire, as well as the medals hung about their necks.

Such trivial proceedings of ceremony have always had good effects with favages, and conquefts made by thefe means have always been the moft lafting.

ARTICLE XVII.

On furveying the iflands, coafts, and promontories, under Ruffian fubjection, you muft, befides the prefcribed information, acquaint yourfelf as accurately as poffible with the number of male inhabitants in fuch places, and begin collecting tribute from thefe people; but, in doing this, you are forbidden to ufe force, or even to revenge incivilities from favages; on the contrary, abftain as much as poffible from Manflaughter, even if they are fo bold as to attack you, as the iflanders of Alkutan, Oomnak, Oonalga, Accoon, and others, have often done to the Ruffian hunters without any provocation; in fuch cafes, remonftrate with them through your interpreters; tell them, that they unreafonably attack fuch as wifh to poffefs their friendfhip; promife and give them fmall prefents; but order all your men to fhew themfelves to as much advantage as poffible from a diftance, to frighten them, and prevent bloodfhed, which in fuch cafes is almoft unavoidable; explain to them, that, if they will not liften to your kind behaviour, you are provided with fuch terrible arms as at once will kill numbers of them, and which you will be compelled to employ if they will not be quiet; for it is impoffible that there fhould be any other reafon for their uncivil and unfriendly behaviour to Europeans, than fuperfluous precaution and fear on their fide; and it is too often the fault of the adventurers, when they attack thefe people with fire and fword, and bring them to a kind of defpair; on the contrary, humane and friendly behaviour keeps them quiet; it is, therefore, ftrongly recommended to you to proceed with them in this mild manner, and not to change your conduct till open and unavoidable danger compel you to fhed blood; keep yourfelf in conftant readinefs, however; employing your arms only to frighten, and not to deftroy, thefe unhappy creatures, endeavouring rather to take one of them alive; and fuch prifoner you may carefs, make him prefents, hang a medal about his neck, explaining to him, that by this you make him your friend, and will know him when he

comes

comes to you again; keep him prifoner as fhort a time as poffible; and, when you releafe him, give him neceffaries, and perfuade him to tell his countrymen of your behaviour to him, and that he may return to the fhip with whom he pleafes, without fear; promifing him, then, prefents of inftruments for catching animals, or whatever he likes; and that he will be received in a friendly manner by all your people, if he only fhews the medal about his neck.

When fuch perfon comes to your fhip with others, tell him, through interpreters, that the fame arms which were before fo alarming to them, will, if they choofe it, be turned into harmlefs thunder, and ferve as a mark of joy for the return of their friends.

You may then prefent them with fuch things as are agreeable to them; treat them with brandy, fugar, or tobacco, which moft of them are fond of; give them likewife traps and gins, fmall copper kettles, knives, needles, and nets, telling them the ufe of what they know not; and defire them to bring you furs, oil of animals, fifh, or what they have; mentioning alfo, that when others come with fuch things, they will receive what they like. Having made them, by thefe means, defirous of vifiting you, you lay a foundation for future collection of tribute; make them incline to trade, to be induftrious in hunting, and more fociable; and thus you will fulfil a principal point of your commiffion, to the glory of Her Majefty, and your own honour.

A R T I C L E XVIII.

Sailing along the above-mentioned iflands, coafts, and promontories, which you are to defcribe, when you come to Cape St. Elias, you may there, in Her Imperial Majefty's name, declare yourfelf Captain of the Firft Rank; and having made on this Cape fuch obfervations as are enjoined for other places, if on your return, about the ifland Oonemak, or the point of Alakfa, fuch weather fhould fet in as to render it unadvifable to keep the fea on account of an approaching winter, you may fearch for wintering on the iflands of Oonalafhka, the bay named by Captain-Lieutenant Levafheff the harbour of St. Paul's, or lie in the ifland Oonemak, in the found oppofite Alakfa, at one verft and a half diftant from Alakfa; and if not this, then go into any of the bays on the coaft of Alakfa, to the eaft or weft, where, upon Captain Krenitzin's affurance, many fine bays may be found within 150 verfts. There, choofing a fafe and proper wintering place, begin immediately to build one or more huts on fhore; ufe all manner of precaution againft the fcurvy, that you may not fuffer as Captain Krenitzin did in wintering on this coaft, who loft about fixty men in this diforder, and was reduced fo low, that, had not Captain-Lieutenant Levafheff come with his people to his affiftance, he would not have had people enow to

manage

manage his ship. For your service on shore, you may take some guns, cartridges, and small shot; for the dreadful example of attacks of islanders upon the Russian adventurers, which they tried also upon Captain Krenitzin when he wintered at Oonemak, must make you as cautious as Captain Krenitzin was, but chiefly against their night-attacks; he had four posts for night watches; had guns and small arms fired at stipulated times every few minutes, to frighten the savages, who tried more than once to overcome the guard, and kill him, with all his people. You must also endeavour, for your security, by fair means and presents, to get the American chiefs to give you some of their children as hostages; to whom you will behave in a friendly manner; but do not take too many of them, that they may not incumber you, particularly if provisions should run short. It is true, their parents used to bring them victuals; but it may happen that they will delay sometimes, and then you must feed them on your own stock. You must order your Command, that such as have been lately wounded, or have some internal disorder, or such as have even long ago had the venereal disease, should not eat whale's flesh; for the wounds will open again, and the venereal disorder will be renewed within three days, as may be seen in Captain Krenitzin's journal.

When you are on the island Oonalashka, endeavour to describe the inhabitants of it more accurately, and enquire of their migrations or origin; why they call themselves Cogolach, as those of the island of Oomnak call themselves Kigigoos, and those of Alaska Cartagaeguk; for the name of Aleutes given to these islanders by the Pilot Nevotshikoff, was taken by him from the name of the islands lying near Kamtshatka. Also, when on the island Oonemak, look (for curiosity's sake) to see whether the wooden cross with a copper crucifix fixed in it, erected by Captain Krenitzin near his winter mansion, be still existing. In a cut in this cross, look for a paper left by him; it will serve you in your intercourse with the islanders.

ARTICLE XIX.

If, during your navigation, it should be necessary to repair your own ship, or if any accident should render it unserviceable, then you must go on board the ship commanded by the second. Taking yourself the command, pursue in it your voyage and observations; in like manner, if the same should happen to the ship of your second, take him and his complement of men on board your own ship; for this reason, the officer commanding the said ship shall be enjoined in his particular instructions never to separate, or stay behind you, excepting at small distance, or by your express order; and that, should this happen in a storm, he shall endeavour as

soon

soon as poffible to rejoin. For greater fecurity, you muft fix frequent rendezvous, that, in cafe of feparation, you may more conveniently join ; and you muft fix night and day fignals for different accidents during your voyage. Should illnefs or other caufes prevent your doing your duty, your fecond is to take the command, and fulfil the tenor of your Inftructions, of which he fhall have on board his veffel a copy figned by you, which you are to give him at your failing from Ochotfk.

A R T I C L E XX.

As it frequently happens in thofe feas, that in the month of October heavy fogs appear, which make it almoft impoffible to fail without danger of lofing yourfelf, as it happened in 1767 in Captain Krenitzin's Expedition with all the fhips, and particularly to the fhip commanded by the Sturman Duding, which was wrecked on the 7th Kuril ifland, called Siafhkuta, where not only the fhip, but almoft all the crew were loft ; you, therefore, and the Commander of the fecond veffel, muft keep a good look out, particularly in unknown places, that no misfortune may happen to the fhip or to yourfelf; which will be a lofs to the Treafury, and a hindrance to Her Majefty's intention.

A R T I C L E XXI.

In all that relates to the fervice of Her Imperial Majefty, you are to conduct yourfelf as a good and experienced officer; and, as well as your fubalterns, endeavour to deferve the graces received, and future promifes ; for this reafon you are to give your fubalterns, whenever you employ them on feparate fervices, clear and determined inftructions, agreeing with the general inftructions given to you ; and oblige them thereby, as you yourfelf are obliged, to be refponfible for faults and omiffions, made purpofely or through neglect.

A R T I C L E XXII.

Having finifhed your enquiries about the iflands, &c. in a good time of the year, or if the ftate of your crew, veffels, and provifions, make it advifeable to hold out another year in thofe feas, then you may take your courfe direct to Bering's Straits, to perfect the knowledge that you will have of the Tfhutfki coaft, and try if you can get by fea to the bay Tfhaoon, or the river Kovima, if by your firft expedition to the Kovima you fhall not have acquired fuch perfect knowledge,

3

that

that all further trials may be ufelefs.　But fhould you find the paffage to the Kovima in large veffels impracticable, then you may, when all that is prefcribed to be
done in the Eaftern Ocean, and about America, is accomplifhed, reach a harbour
on the Tfhutfki coaft; and, if it promife fuccefs, land there with a neceffary
number of men and inftruments, giving orders to the commanding officer that
remains in the fhips how long they are to wait for you (if you think it advifeable to
keep them there), and that afterwards they fhall return to Kamtfhatka, or Ochotfk,
where they are to expect your further orders.　If the fea fhould be free from ice
along fhore, you may take fome row-boats from the veffels, giving, however, fome
to the fhips; or build there baidars from materials prepared before; by the help of
which, fometimes by land and fometimes by water, you will try to get round to the
river Kovima, laying down your route upon the chart, and making neceffary obfervations, chiefly for determining what is not yet fettled on the charts.

But if, after thefe trials to the north, you return yourfelf in your frigates towards Kamtfhatka or Ochotfk, you may endeavour to make your return as ufeful
to geography as poffible, coafting round the bay of Anadir, or touching at fuch
iflands as you could not fetch in your firft voyage.

ARTICLE XXIII.

At your arrival at the port of Kamtfhatka, and afterwards at Ochotfk, you
have to return the Sturmen, Coffacks, Interpreters, and Kamtfhadals, to their
refpective commands and places of abode in the government of Irkutfk, with
written certificates of their behaviour, and recommendations for what each
deferves.

You will alfo deliver your veffels, ftores, ammunition, and provifions remaining,
by fpecification, againft receipt, to the Commander of Ochotfk; and if you can
fpare fome inftruments, without hindrance to the obfervations you may make
on your return, you may alfo deliver fuch againft receipt for the future navigation
of Ochotfk.

ARTICLE XXIV.

Having thus finifhed your Chief Expedition, and collected your Command
that is to return to St. Peterfburg, you are to make preparations without delay
for your return, which make as ufeful as poffible to the geography of the different parts of Siberia.　With this view, you may fend fome of your fubalterns
with

with proper inftruments on a different route; they might go with Mr. Patrin up the river Viluie, and from thence over the river Neizfhnoi or Pod-Kaminoi Tongufka, to the river Jenefei, to furvey the natural curiofities unexplored in thofe parts. They would do fervice to geography if they could obtain fome knowledge of the advanced point which ftretches farther than any other part of Siberia towards the Pole, between the rivers Olenek and Jenefei, more efpecially between the Katanga and Taimura; it may be, befides, that you may have oppor-tunities of determining or rectifying the longitude and latitude of remarkable places not fpecified in the lift annexed; you will likewife furvey remarkable rivers, which is not to be neglected.

A R T I C L E XXV.

To conclude this Inftruction, approved by Her Majefty, that nothing may be wanting to encourage your zeal, Her Imperial Majefty has been pleafed to order the important truft to be laid on you, of making alterations in what is pre-fcribed in the Articles, according to your judgment and circumftances, with the common confent of your officers; chiefly, however, when undoubted advantages may arife therefrom to the Expedition, for the good of the fervice and the Em-pire. This great truft will, doubtlefs, raife in your heart and thoughts a noble emulation of fuch great men as have to their honour been employed in like fer-vices as you are charged with; and will excite you to think only how you fhall begin with zeal, purfue with good fenfe, and end with honour, this important charge.

A D D I T I O N A L A R T I C L E

To the Inftructions of Captain-Lieutenant Billings.

On the chart oppofite the river Kovima, to the north from Bear Iflands, is marked the coaft, which ftretches as a continuation of the Continent of America. This has been adopted from a chart fent by Governor Tfhetchirin in the year 1764. A fergeant Andreeff faw from the laft of the Bear Iflands, at a very great diftance, what they thought a large ifland, toward which they went with dog fledges on the ice, but did not arrive at it by twenty verfts; they found frefh foot-fteps of a great number of people who had been that way in rein-deer fledges; but they, being few in number, returned to the Kovima. No later account of the large ifland, or continent, has been received; it is therefore thought neceffary to make you obferve this; as you will be on the river Kovima, and not far from

thence,

thence, it would be useful if you could possibly survey and describe, or at least get nearer accounts of the circumstances of this land; whether it be an island or part of the continent of America; if there be inhabitants, and how great a number; in general, make all such inquiries as are prescribed concerning newly-discovered lands. This, however, is so recommended to your observation, as not to intrude on your chief occupation.

No. VI.

INSTRUCTIONS

FOR

THE NATURALIST, MR. PATRIN,

Who is ordered to accompany the Expedition deſtined for the Kovima and the Frozen Ocean.

[*The Original in French.*]

H<small>ER</small> Imperial Majeſty having been graciouſly pleaſed to appoint you in quality of Naturaliſt, on a voyage of diſcovery about to be undertaken under the Command of Captain-Lieutenant Billings toward the Kovima, the Eaſtern and Frozen Ocean; every exertion is expected from you, which your honour, and your zeal for the ſciences which you profeſs, and for the ſervice you are engaged in, can prompt: the more ſo, as Her Majeſty, for your encouragement, has been pleaſed to give you one rank more than you now hold in the ſervice of the mines, to take place from the day on which you join the Expedition; likewiſe a ſum of   rubles to defray the expences of your equipment; and double pay during the term of the Expedition; in which you will certainly have opportunities of making diſcoveries, and rendering ſervices, which will entitle you to the further protection of Her Imperial Majeſty.

In order to give you a full inſight into what is expected from you, Her Imperial Majeſty has been graciouſly pleaſed to approve the following articles, to ſerve for your inſtruction.

ARTICLE I.

Upon the arrival of Captain-Lieutenant Billings at Irkutſk, you will paſs from the ſervice that you are now employed in, to the Expedition under his command,

with

with which you are to continue fo long as it lafts, and with which you will return to St. Peterfburg; where you will give up your journals, obfervations, and collections, together with fuch fpecimens of, natural hiftory as you may have collected, to the department which Her Imperial Majefty will name for their reception.

ARTICLE II.

You are to follow the Commander of the Expedition in all his journies by land and voyages by fea, beyond the river Lena; and you will affiduoufly obferve all that is prefcribed in thefe inftructions; particularly in thofe parts of Siberia, as well as coafts and iflands, which have never been vifited by naturalifts; fuch as the banks of the Kovima, the coafts of the Frozen Ocean of the Pacific, and Kamtfhatka, and the iflands you will there touch at. You will keep an exact journal of the voyage, together with a topographical defcription of the countries that you are to pafs through, their rivers, lakes, and mountains; the productions in the three kingdoms of nature, and the inhabitants. You will alfo make meteorological obfervations, and remarks upon the feveral properties of the countries that you may vifit, from the beft intelligence you can collect.

ARTICLE III.

You will defcribe in a very particular manner the extent, connexion, and direction, of the chain of mountains; their fhapes, fuperfices, declivities, and heights; the rocks or foils of which they are compofed; the ftrata that they contain, and their direction; craters, remains of extinguifhed volcanoes, and fuch as are actually burning. You are to collect fpecimens of all forts of rocks, earths, petrifactions, lava, foffil, remains of animals, minerals, falts, and fulphurs; carefully numbering them, and noting the fpot where found; alfo collect all remarkable ftones and pebbles brought down by rivers, or thrown up by the fea, as well as fuch as may be in ufe by the inhabitants.

You will defcribe the furface of the country, its irregularities, and the layers of foil found at different depths; the fituation of the country, whether low or elevated; woods and underwoods, animals, birds, marfhes, lakes, rivers great and fmall, the nature of the waters, efpecially if they appear to have any particular qualities, the fifh found in them, and every other remarkable production.

ARTICLE

ARTICLE IV.

With regard to the people that you may visit, you will observe their dispositions and different corporeal qualifications; their government, manners, industry, ceremonies, and superstitions religious or profane; their traditions, education, and manner of treating their women; useful plants, medicines, and dyes; food, and manner of preparing it; habitations, utensils, carriages, and vessels; manner of life and economy; their modes of hunting, fishing, making war, and treatment of domestic animals; likewise languages, of which you will collect vocabularies, according to the plan sent with the Expedition, marking the pronunciation according to the Latin orthography. You will also try to procure the dresses, ornaments, instruments, and arms of these people, or cause them to be drawn. You will likewise make descriptions of tombs and other monuments of antiquity.

ARTICLE V.

You will particularly attend to trees, shrubs, land and water plants; preserving as many specimens as possible, particularly of any that are extraordinary or new; and you will employ your leisure time in making complete descriptions of such specimens; noting the season of their growth, flowering, and maturity. You will lose no opportunity of remarking most minutely such as may be of benefit to society, and which you may discover to be of use as food for man or beast, or applied as a remedy for any disorder; the manner of preparing dyes, stuffs, or skins. You will collect specimens of woods, barks, gums, resins, remarkable fruits, bulbs, and roots; as also every thing that may be cultivated in the gardens of Europe, noting the provincial and natural names.

ARTICLE VI.

You will collect, and cause to be stuffed or otherwise preserved, all extraordinary quadrupeds, birds, fish, amphibious animals, insects, shell-fish, or zoophytes; observing as closely as possible their habits, food, propagation, sounds, migrations, and habitations, as well as the mode of catching them, with the instruments and stratagems made use of for that purpose. You will also collect as many species of birds' eggs as possible. Quadrupeds and birds of different genders and ages are to be stuffed; fish, amphibious animals, and zoophytes, to be preserved in spirits of wine; insects, shells, and dried productions, fixed or packed up in cases made for that purpose.

ARTICLE

A R T I C L E VII.

Meteorological obfervations, particularly thofe with the thermometer and ba-rometer, demand your ftri&teft attention; but moft fo in the places where you may winter or ftay any time. You will form tables of thefe obfervations in the ufual manner, noticing all remarkable phenomena, fuch as Parhelii, Aurora Boreales, and their concomitant circumftances; obferve the congelation of mercury in dif-ferent manners by natural and artificial cold; and determine by the fpirit thermo-meter the true point of congelation. The altitude of different mountains may be determined by correfponding barometrical heights.

Although the predominant or variable winds, tides with their changes and di-rections, currents, and other nautical occurrences, are the more particular bufinefs of the Commander, you will not negle&t to make fuch obfervations as you can, and note them in your journal.

A R T I C L E VIII.

You will inform yourfelf of all national illneffes, efpecially endemic or epi-demic, which exift in particular latitudes, or among particular nations; the dif-tempers of domeftic animals and horned cattle; and the remedies moft in ufe to prevent or eafe them.

A R T I C L E IX.

You will be careful in preferving the natural curiofities that you may colle&t, numbering them, and keeping a catalogue containing the places where found, with defcriptions and other obfervations; or all this may be expreffed on each label. The ftuffed birds or animals muft be carefully dried, and fmoked with fulphur, before they are packed up; the boxes or packages dried and fmoked in like man-ner, and the cafes covered with pitch and with leather. To every article likely to be fpoiled by infe&ts or damp, particular attention muft be paid. When the Commander makes his reports, you will alfo fend your obfervations, and fuch colle&tions as are convenient; the others are to remain in your cuftody till your return to St. Peterfburg.

ARTICLE

A R T I C L E X.

You may require from the Commander of the Expedition fuch affiftance of men, horfes, inftruments, and money, as may be neceffary for your phyfical operations; and when your prefence is not neceffary with the Expedition, you may make excurfions, with the Commander's confent, into the neighbouring country, where you may expect to meet with objects worthy of your remarks, either phyfical or hiftorical. You will receive every affiftance for this purpofe from the Commander of the Expedition; and the draftfman may accompany you if he be not employed on more important bufinefs.

(Signed)   P. S. PALLAS.

No. VII.

No. VII.

EXTRACTS

AND

SUPPLEMENTARY OBSERVATIONS.

The following Remark was made in Captain Billings's Journal, by his order, while at Oonalaſhka in 1790, on the Iſland Sithanak.

"In conſequence of complaints made to me in form, upon my firſt arrival at " Ochotſk, by ſeveral people who were ſent by Government to collect tribute of " the Aleutan iſlanders, againſt the hunters, for cruelties to the natives, I repre- " ſented the ſame, and received a private Mandate from Her Imperial Majeſty, " ordering me to inſpect the behaviour of the merchants and hunters in theſe parts. " I have, in conſequence, made it my buſineſs at Sithanak and Oonalaſhka to make " enquiries into the treatment which the natives receive from theſe people; and " have been, as well as every gentleman on board, an eye-witneſs of the abject " ſtate of ſlavery in which theſe unfortunate iſlanders live under the Promyſh- " lenicks (hunters). The company now at Oonalaſhka conſiſts of twelve Ruſſians " and one Kamtſhadal (their veſſel is in the ſtraits of Alakſa or thereabouts). Theſe " people employ all the men of Oonalaſhka and Sithanak in the chaſe, taking the " fruits of their labour to themſelves, and not even allowing the natives neceſſary " clothing. There is, therefore, no name ſo dreadful to them as that of Peredof- " ſhick (the leader of a gang of hunters). Upon the arrival of their veſſel at any " place where they purpoſe making a ſtay, they haul her on ſhore; immediately " ſend the natives out on the chaſe, even to the fartheſt of Shumagin's iſlands; " and then take by force the youngeſt and moſt handſome of the women for their " companions.

" If another veſſel arrives, they unite their companies, or elſe the ſtronger " party takes the natives from the weaker. They inflict on the natives what " puniſhments they pleaſe, and are never at a loſs to invent a cauſe."

TRANSLA-

TRANSLATION

Of a part of the Journal of one of our Ruſſian Officers while at Oonalaſhka in 1790.

" The company of hunters now here make their boaſt that they clothe and feed
" the iſlanders; which they do in the following manner : The natives, being under
" their controul, are ſent out in parties to chaſe ſea animals and catch fiſh. The
" produce of the chaſe is delivered into the Company's ſtock, out of which the
" natives receive an allowance. Such of the inhabitants as are too infirm or too
" young to be ſent out on aquatic excurſions, are employed in domeſtic drudgery,
" and digging edible roots; while the women are occupied in making and mending
" clothing from the inferior ſkins of animals and of birds."

" The hunters were accuſtomed to act as follows : Upon the arrival of any
" veſſel at an inhabited iſland, the Peredoſſhik ſent an armed boat to the habitations,
" to take from the natives all the furs and valuable articles that they poſſeſſed;
" and, if the leaſt oppoſition was made, they were ſilenced by the muſkets of the
" hunters. Wives were taken from their huſbands, and daughters from their
" mothers; indeed the barbarity of their ſubduers to the crown of Ruſſia is not
" to be deſcribed. They uſed not unfrequently to place the men cloſe together,
" and try through how many the ball of their rifle-barrelled muſket would paſs *.
" Nor were the hunters more kind to their own brethren; for if two parties in
" different intereſts met, they fought together for the poſſeſſion of the natives, or
" formed themſelves into one company."

A Bird of the Auk kind caught at Oonalaſhka.

Bill orange colour, very little curved; both mandibles tipped and edged with
black; the noſtrils long and narrow, running parallel with the mouth; an elevation upon the noſtrils of a light green colour, edged with black. The feathers
commence at the baſe of the bill, and are of a dark aſh, which is the colour of the
head and neck. From the upper part of the eye, along the head, to the back of the
neck, is a row of fine white ſatin feathers; and another row, broader and ſhorter,
leads from the corners of the mouth: The eye of a pale yellow, the pupil being
ſmall and of a very dark blue. The back, ſcapulars, coverts of the wing, and tail,

* Gregory Shelikoff has been charged with this act of cruelty; and I have reaſon to believe it, from
the teſtimony of ſeveral Ruſſians at Ochotſk, corroborated by ſome of the natives of this iſland.

8

are

are dark, with a paler edging; primaries something lighter; throat a light colour; breast and belly a dirty white; the fore part of the legs of a livid colour; the hind part, web, and claws, black, with three toes. It resides about the rocks and coast of Kamtfhatka, and upon all the Aleutan iflands, and is about the fize of a black-bird.

Fifh caught at Oonalafhka, March 23, 1792.

Angling among the rocks, the hook baited with the common edible mufcle, I caught a fifh called by the Ruffian hunters *terpug* (*rafp*). It is fixteen inches long, and fhaped like a mackerel. The head of a dark olive, with fcarlet fpots. Behind each eye, on the top of the head, is a palmated flefhy creft half an inch long, and one-eighth broad. It has five branchioftigous rays, prominent and ftrong; thefe and the lower part of the head are of a lively fcarlet. The colour of the body of the fifh is dark olive, with blotches of fcarlet, and a dull red; two dorfal fins fpotted in the fame manner, and united at the extremities; both rounded; the firft confifts of twenty rays, the fecond of twenty-two. The pectorals large and rounded, eighteen rays, fpotted at the dorfal, but edged with fcarlet, as is alfo the anal fin, confifting of twenty-two rays; ventral five rays; tail rounded; breaft and throat a lively fcarlet. On each fide of the breaft is a line of fmall dots, reaching between the ventral and pectorals, turning up to the latter, and extending in a ftrait line to the tail, very high on the back; a fimilar line encircles the dorfal fin, there is another half an inch below it, and one near the ventrals. The flefh, gills, and infide of the mouth, are of a lively light blue, inclining to green; when boiled it turns white, but the bone retains fomewhat of this colour. The fcales are fmall and rough, whence it derives its name.

The fame day I caught another fifh, about feven inches long; head large, but fhort; the fides of the bony plates and head replete with fmall pits; large mouth, with fharp clofe-fet flender teeth. The dorfal fin reaches from the hind part of the head to near the tail, which is rounded. The fifh is very fmooth; its colour a dark olive marbled with dufky green, edged with a dull red, forming broad bars that crofs the lateral line, which is ftraight.

A very black fifh refembling a carp I frequently caught lurking under ftones; as alfo the father-lafher. The armed bull-head alfo was caught in our net, and the fpotted blenny.

[H]   I took

I took one fifh which adhered very faft to a rock by means of a fucker on its belly. It is very fhort and thick, and the flefh flabby; but it boiled firm.

I alfo found a fifh lying dead on the beach, about five feet long, round, and fhaped like an eel, with a large mouth, and very fharp teeth.

The other kinds of fifh are, halibut, cod, thornback, and feveral fpecies of falmon.

THE END.

Printed by A. Strahan,
Printers-Street.